INTERNATIONAL CENTRE FOR MECHANICAL SCIENCES

COURSES AND LECTURES - No. 103

PETER W. LIKINS
ROBERT E. ROBERSON
UNIVERSITY OF CALIFORNIA
JENS WITTENBURG
UNIVERSITY OF HANNOVER

DYNAMICS OF FLEXIBLE SPACECRAFT

DEPARTMENT OF GENERAL MECHANICS
COURSE HELD IN DUBROVNIK
SEPTEMBER 1971

UDINE 1971

SPRINGER-VERLAG WIEN GMBH

Originally published by Springer-Verlag Wien New York in 1972

ISBN 978-3-211-81199-3 ISBN 978-3-7091-2908-1 (eBook)

DOI 10.1007/978-3-7091-2908-1

P R E F A C E

At the suggestion of Professor Sobrero of CISM, I organized a series of lectures to be presented by me and several colleagues at Dubrovnik in September 1971, under the joint auspice of CISM and the University of Zagreb. For his encouragement and support, I wish to immediately express my thanks.

The lectures were organized in two series, and three hours of lectures were presented in each series each day during 13 - 17 September. This book contains the lectures of the second series, given by Undersigned, by Professor P.W. Likins, and by Dr. W. Wittenburg. Each series was devoted to one aspect of special current importance relating to the rotational behaviour of spacecraft.

The subject of this second series was the dynamics of flexible rotating spacecraft. This is a topic of considerable current interest to rotational dynamics as a science, as well as to its technological application area of rotating spacecraft. We have attempted to describe here the two major approaches to the problem: first, the approach through linearly elastic dynamical equations, generalized from the traditional structural dynamical equations by reference to rotating bases; second, the approach through the dynamics of a discrete set of interconnected, individually rigid bodies. Each has its own domain of applicability.

Professor Likins prepared and presented Lectures 1 - 9; Dr. Wittenburg, Lectures 11 and 14, and I the remaining lectures.

Udine, September 1971

Peter W. Likins

Robert E. Roberson

Jens Wittenburg

Introductory remarks

As I have remarked previously, spacecraft problems have been responsible for a resurgence and growth of the dynamical theory of rotating systems. For many purposes during the last two decades it has been possible to model the spacecraft as a rigid bodies or gyrostats, or perhaps simple two-body systems. However, even from the earliest days of real satellites, cases have been known where non-rigid characteristics have dominated the dynamical behavior.

Within the last few years, elastic deformations have become of increasingly great importance in both spin-stabilized and passively stabilized systems, because the elastic behavior can be central to the stability of the desired state of motion. Furthermore, even in actively controlled systems elastic behavior has become increasingly important as the size of the proposed structures increases and the accuracy of pointing control becomes greater for certain applications.

We no longer can afford to focus solely on the rigid body aspects of spacecraft rotation, but must begin to consider its elastic behavior as well. The resulting class of problems lie at an interesting triple point between classical rigid body dynamics, the theory of structures, and control theory, presenting new facets to the disciplines now familiar in each of these fields.

This lecture series is intended to convey some of the flavor of this relatively new aspect of spacecraft dynamics. It is not a definitive treatment, for such is not yet possible. The field is growing and changing, and currently represents a forefront research. It is felt, nevertheless, that the material presented here is a sound foundation on the basis of which the listener can further develop his own interests.

1. (*) Mathematical modeling of spacecraft

A lecture series devoted to the Dynamics of Flexible Spacecraft is concerned not with a single problem but with a family of related problems. Accordingly, there is not a single correct approach to solution, but a spectrum of methods to be applied to a family of spacecraft idealizations. Although analysts will differ in their preference for various ways of formulating equations, one man choosing Lagrange's equations, a second preferring Hamilton's principle, and a third relying upon a Newton-Euler formulation, these differences are much less fundamental

(*) Lectures 1 through 9 by Likins are based largely on work sponsored by NASA, through either the Jet Propulsion Laboratory or Marshall Space Flight Center. Nomenclature for these lectures is listed following section 9, commencing on page 93.

than the initial choice of a mathematical model of the vehicle.

The spacecraft mathematical model consists firstly of an idealization and mathematical description of the physical system (this we might call the mechanical model), and secondly of a mathematical statement of the motions which can be experienced by the idealized spacecraft (this then becomes the kinematical model). A representative list of options to be considered in each of these modeling decisions might be drawn up as follows:

A. Mechanical models

(a) Rigid body

(b) Elastic or viscoelastic continuum

(c) Collection of elastic elements ("finite elements") interconnected at nodal rigid bodies or particles

(d) Collection of interconnected substructures, each of which is modeled as in a), b), or c).

B. Kinematical models

(a) Unrestricted coordinates of the mechanical model

(b) Coordinates restricted to allow only "small" deformations

(c) Coordinates partially prescribed by interpolation functions

(d) Combinations of a), b), and c).

Although mechanical model Aa), the rigid body, has served to represent most early spacecraft after launch, and mechanical model Ab), the elastic continuum, has proven useful in the representation of launch vehicles modeled as elastic beams, still these basic models must be said to have quite limited utility in application to modern space vehicles. Model Ac) is appropriate for any spacecraft which may be idealized as linearly elastic and subject to small deformations, but this is still in a modern context quite a restricted class of vehicles. In most current applications one must resort to model Ad), which involves the subdivision of a spacecraft into a collection of substructures, and independent idealization of the individual substructures. In this fashion one might accomodate an actively controlled scanning antenna of great flexibility mounted on a spacecraft frame which is itself essentially rigid, or one might connect two or more rigid bodies or two or more elastic bodies. The combinations are many and varied, as are the vehicles to be analyzed.

In order to develop a rationale for adopting a particular idealized mechanical model, one must give some thought to the anticipated kinematical model. A rigid body (model Aa)) is, of course, fully characterized in its motions by six scalar coordinates, and six second-order (or twelve first order) ordinary differential equations will always suffice to predict its motions. The number of independent coordinates increases for a collection of rigid bodies in a manner established by the constraints among the bodies, being always no greater than $6n$. In working with such systems, it is customary to deal with some collection of scalar coordinates each of which describes a kinematical property of a particular rigid body of the system. Such coordinates may be characterized as discrete, in contrast to the distributed coordinates to be described next.

A continuum (model Ab) must be characterized kinematically not by scalar coordinates depending only on time, but by scalar functions of space and time; the equations of motion must be partial differential equations. Even for the simplest continuous model of a spacecraft (e.g., a uniform elastic beam), one normally finds it advantageous to replace the partial differential equation by a large but finite number of ordinary differential equations, expressed in terms of coordinates each of which describes a motion or deformation in which the entire vehicle (or substructure) participates; these are called distributed coordinates. The substitution of a finite number of ordinary differential equations for a partial equation evidently involves an approximation, since the continuous system originally postulated could be described kinematically only by an infinite number of scalar coordinates varying only with time. In the simplest case (e.g., a uniform elastic beam vibrating freely about a state of rest in inertial space), the transition from partial to ordinary differential equations is accomplished formally by representing the solution to the unknown function in the partial differential equation as a product of two functions, one of which depends only on spatial coordinates and the other only on time. The latter then provides the unknowns in the ordinary differential equations, while the former provides a shape function which describes the spatial distribution of motion (or deformation) for a unit value of the latter. The expeditious decision to work with a finite number of distributed coordinates is then implemented by simply truncating the infinity of coordinates formally obtained and electing to proceed with a smaller number judged to be representative of the salient features of the system dynamics. This is not a rigorous step mathematically, but it need not be inconsistent with the level of validity of the mathematical model originally adopted for the spacecraft (no real space vehicle is a uniform, homogeneous, isotropic, elastic beam).

In most realistic situations, it is impractical to begin with a partial differentail equation of motion for a material continuum model of a space vehicle, and alternatives must be found.

It is a common practice among those attempting to use a continuous model of a spacecraft or complex spacecraft component to avoid the partial differential equation from the outset, relying upon an initial formulation in terms of a finite number of distributed coordinates which the analyst judges to be adequate to represent all dynamically significant structural deformations. Note that when one adopts this practice he begins with a continuous mechanical model but immediately imposes a kinematical model which restricts the number of degrees of freedom of the system. It then becomes equivalent to the adoption in the first place of a mechanical model having a limited number of degrees of freedom, as in model Ac).

Probably the mechanical model most commonly adopted for the representation of a complex flexible substructure of a spacecraft is model Ac), which idealizes an elastic body as a collection of nodal bodies (either particles or rigid bodies) interconnected by elastic members, often called <u>finite elements</u>. The deformable finite elements may be massless, or mass may be distributed throughout each deformable element; in the latter case the nodal bodies might be massless.

Since small deformation theory is generally employed in analyzing the deformations of the elastic elements in mechanical model Ac), one must generally sacrifice the generality of the unrestricted coordinates in kinematical model Ba). If the finite elements have been idealized as massless, then kinematical model Bb) is appropriate, so that nodal bodies of a given substructure are permitted to experience only small relative motions (perhaps in conjunction with large common motions in inertial space). If the continuous finite elements are idealized as having distributed mass, then the mechanical model again has an infinite number of degrees of freedom. The number of coordinates is reduced by employing kinematical model Bc), for which interpolation functions are introduced to provide the deformations within a finite element explicitly in terms of the relative motions of the nodal bodies.

It is not possible to consider in intelligible detail within the limits of this lecture series the dynamic analysis of each of the mathematical models listed, although the engineer responsible for spacecraft analysis really should have some knowledge of the advantages of each. Rather than provide a superficial survey of all potentially valuable procedures, we have elected to examine in depth a limited number of approaches. Specifically, we shall explore initially the formulation of equations of motion of a flexible substructure modeled as interconnected nodal bodies as in Ac), when attached to spacecraft components modeled as rigid bodies as in Aa); this combination will involve a combination of discrete and distributed coordinates, so it is called a <u>hybrid coordinate</u> formulation.(*) After deriving the appropriate equations for this special case (and indicating briefly how one might approach the more general case involving the coupling of flexible substructures), we shall explore some of the results recently obtained by means of hybrid coordinate analysis. The second half of our lecture series will deal

(*) The material immediately following is drawn from Ref. 18 of the Bibliography.

with the formulation and use of equations of motion of mathematical models consisting of collections of interconnected rigid bodies.

2. Appendage idealization

Any portion of a vehicle which can reasonably be idealized as linearly elastic and for which "small" oscillatory deformations may be anticipated (perhaps in combination with large steady-state deformations) is called a flexible appendage.

A flexible appendage is idealized as a finite collection of ε numbered structural elements, with element number s having n_s points of contact in common with neighboring elements or a supporting rigid body, $s = 1,\ldots,\varepsilon$. Each contact point is called a node, and at each of the n nodes there may be located the mass center of a rigid body (called a nodal body), but the elastic structural elements may also have distributed mass.

Figure 1 is a schematic representation of an appendage (enclosed by dashed lines) attached to a rigid body b of a spacecraft, which may consist of several interconnected rigid bodies and flexible appendages. A typical four-node element of the appendage is shown in three configurations of interest: i) prior to structural deformation, ii) subsequent to steady-state deformation, induced perhaps by spin, and iii) in an excited state,

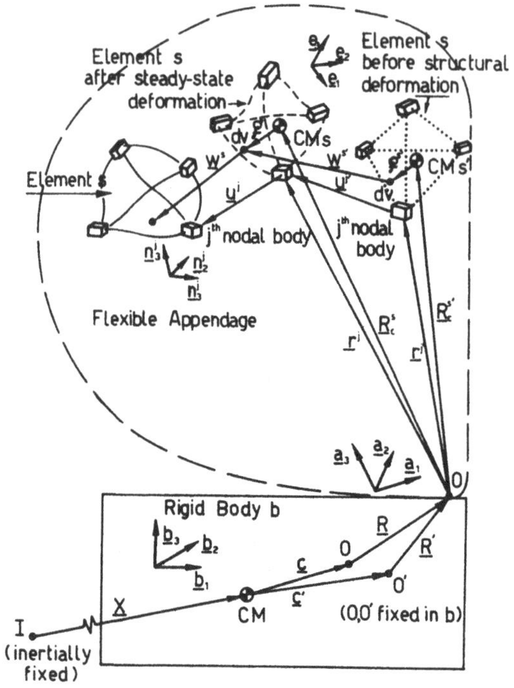

Fig. 1

experiencing both oscillatory deformations and steady state deformations.

The point Q of body b is selected as an appendage attachment point. The dextral, orthogonal unit vectors $\underline{b}_1$, $\underline{b}_2$, $\underline{b}_3$ are fixed relative to b, and the dextral, orthogonal unit vectors $\underline{a}_1$, $\underline{a}_2$, $\underline{a}_3$ are so defined that the flexible appendage undergoes structural deformations relative to a reference frame α established by point Q and vectors $\underline{a}_1$, $\underline{a}_2$, $\underline{a}_3$. Gross changes in the relative orientation of a and b are permitted, in order to accomodate scanning antennas and such devices; this is accomplish

ed by introducing the time-varying direction cosine matrix $\underline{C}$ relating $\underline{a}_\alpha$ to $\underline{b}_\alpha (\alpha = 1,2,3)$ by

$$\begin{Bmatrix} \underline{a}_1 \\ \underline{a}_2 \\ \underline{a}_3 \end{Bmatrix} = \begin{bmatrix} C_{11} & C_{12} & C_{13} \\ C_{21} & C_{22} & C_{23} \\ C_{31} & C_{32} & C_{33} \end{bmatrix} \begin{Bmatrix} \underline{b}_1 \\ \underline{b}_2 \\ \underline{b}_3 \end{Bmatrix} \quad (1)$$

or, in more compact notation, by

$$\{\underline{a}\} = C\{\underline{b}\} . \quad (2)$$

The equations of motion to follow permit arbitrary motion of b and arbitrary time variation in C , although practical application of the results requires that the inertial angular velocities of **a** and **b** remain in the neighborhood of constant values over some time interval. These angular velocities will not emerge as solutions of equations to be derived here; the complete dynamic simulation must involve equations of motion of the total vehicle and each of its subsystems, as well as differential equations characterizing necessary control laws for automatic control systems, and only the differential equations of appendage deformations are to be developed here.

As shown in Fig. 1, appendage deformations are described in terms of two increments, one steady-state and the other oscillatory. This separation is necessary because in formulating the equations of motion for the small oscillatory deformations of primary interest here one must characterize the elastic properties of the appendage with a stiffness matrix, and the elements of this matrix are influenced by the structural preload associated with steady-state deformations, as induced for example by spin.

The jth nodal body experiences due to steady-state structural deformation the translation $\underline{u}^{j'} = u_\alpha^{j'} \underline{a}_\alpha$ (summation convention) of its mass center, and a rotation characterized by $\beta_1^{j'}$, $\beta_2^{j'}$, $\beta_3^{j'}$, for sequential rotations about axes parallel to $\underline{a}_1$, $\underline{a}_2$, $\underline{a}_3$. The steady-state deformations of a typical element are represented by the function $\underline{w}'$, which is related to the corresponding nodal deformation by the procedures of finite element analysis. The task of solving for the steady-state deformations of appendages on a vehicle with constant angular velocity is mathematically identical to a static deflection problem. Because, at least formally, large deflections and resulting nonlinearities are to be accomodated, this task is not trivial, but it is in this paper assumed accomplished, so that steady-state deformations and structural loads associated with nominal vehicle rotation are assumed known.

Attention is to focus here on the small, time-varying deformations of appendages induced by transient loads or deviations from nominal vehicle motion. The jth nodal body experiences the translation $\underline{u}^j = u_\alpha^j \underline{a}_\alpha$ and the rotation $\underline{\beta}^j = \beta_\alpha^j \underline{a}_\alpha$ (small angle approximation) in addition to the previously described steady-state deformations. The oscillatory part of the deforma tion of a generic element is represented by the vector function $\underline{\omega}$. (Should it become necessary to deal with such deformations for more than one element simultaneously, the notation $\underline{\omega}^s$ is employed for element s). The quantities $\underline{u}^j$, $\underline{\beta}^j (j = 1,\ldots,n)$ and $\underline{\omega}^s (s = 1,\ldots,E)$ or their scalar components are referred to as variational deformations.

For convenience in calculations it is often desirable to introduce for each finite element in its steady-state condition a local coordinate system, by introducing a set of dextral, orthogonal unit vectors $\underline{e}_1$, $\underline{e}_2$, $\underline{e}_3$, an origin Q and a corresponding set of axes ξ, η, ζ . (Superscripts are appended to each of these symbols should it become necessary to distinguish the particular element.) The local vector basis is then related to the appendage global vector basis $\underline{a}_1, \underline{a}_2, \underline{a}_3$ by a constant direction cosine matrix $\bar{C}$, as in

$$\begin{Bmatrix} \underline{e}_1 \\ \underline{e}_2 \\ \underline{e}_3 \end{Bmatrix} = \bar{C} \begin{Bmatrix} \underline{a}_1 \\ \underline{a}_2 \\ \underline{a}_3 \end{Bmatrix} \quad \text{or} \quad \{\underline{e}\} = \bar{C}\{\underline{a}\} . \tag{3}$$

The vector function $\underline{\omega}$ is most conveniently expressed in terms of local coordinates and the local vector basis; the (3×1) matrix function $\bar{\omega}$ defined by

$$\underline{\omega} = \bar{\omega}_\alpha \underline{e}_\alpha = \{\underline{e}\}^T \underline{\omega} \tag{4}$$

represents $\underline{\omega}$ in the local basis, whereas the (3×1) matrix function $\underline{\omega}$ defined by

$$(5) \qquad \underline{\omega} = \omega_\alpha \underline{a}_\alpha = \{\underline{a}\}^T \omega$$

represents $\underline{\omega}$ in the global basis. Similar notation distinguishes the vector bases of all matrices representing Gibbsian vectors.

An important aspect of the appendage idealization is the assumption to be incorporated in the following section, that the deformations of each finite element can be represented as a function only of the deformations of its nodes, and that the nature of that interpolation function can be imposed a priori.

3. Substructure equations of motion

a) Finite element equations of motion

Having adopted an appendage idealization, one can proceed formally to derive its equations of motion. Since it is the variational nodal deformations $\underline{u}^j$ and $\underline{\beta}^j (j = 1,\ldots,n)$ which represent the appendage unknowns, the equations of motion of the appendage ultimately consist of the $6n$ scalar second order differential equations of motion for the n nodal bodies. The present section, however, has the intermediate objective of providing an expression for the interpolation function relating the variational deformation function $\bar{w}$ of a finite element to the variational deformations at its nodes, and in terms of this relationship providing expressions for the forces and torques applied to the nodal bodies by the adjacent finite elements.

Rather than attempt to work with the infinite number of degrees of freedom of the element as a continuous system, one can avoid introducing any additional degrees of freedom attributable to element mass by assigning to $\underline{w}(\xi,\eta,\zeta)$ a functional structure permitting its representation in terms of the $6N$ scalars definig the translational and rotational displacements due to oscillatory deformations at its N nodes. (*) Although much is left to the discretion of the analyst in choosing an expression for the function $\underline{w}(\xi,\eta,\zeta)$, it is required for present purposes that this expression involve $6N$ scalars $\Gamma_1,\ldots,\Gamma_{6N}$, matching in number the unknown deformational displacements at the N nodes of the element. Typically, polynomials in the Cartesian coordinates ξ,η,ζ are chosen, with $\Gamma_1,\ldots,\Gamma_{6N}$ providing the coefficients. In matrix form, the indicated relationship is written

$$\bar{w} = P\Gamma \tag{6}$$

where $\bar{w}$ is defined by Eq. (4), $\Gamma \triangleq [\Gamma_1 \Gamma_2 \ldots \Gamma_{6N}]^T$, and P is a $(3 \times 6N)$ matrix establishing the assumed structure of the deformational displacement function.

Eq. (6) applies throughout a given finite element, and hence it applies at the element nodes; if the jth node of the

(*) The symbol N_s represents the number of nodes of element s , but the symbol N will be used for a generic element.

element in question, with local coordinates ξ_j, η_j, ζ_j, the nodal displacement $\underline{u}^j$ as represented by the matrix $\bar{u}^j$ in the local basis is from Eq. (6) given by

$$\bar{u}^j = \bar{w}(\xi_j, \eta_j, \zeta_j) = P(\xi_j, \eta_j, \zeta_j)\Gamma \tag{7}$$

and the rotation $\underline{\beta}^j$ is represented in the local basis by the matrix

$$\bar{\beta}^j = \frac{1}{2}\tilde{\nabla}\bar{w}\Big|_{\xi_j, \eta_j, \zeta_j} = \frac{1}{2}\tilde{\nabla}P\Big|_{\xi_j, \eta_j, \zeta_j}\Gamma \tag{8}$$

where

$$\tilde{\nabla} \triangleq \begin{bmatrix} 0 & -\partial/\partial\zeta & \partial/\partial\eta \\ \partial/\partial\zeta & 0 & -\partial/\partial\xi \\ -\partial/\partial\eta & \partial/\partial\xi & 0 \end{bmatrix}$$

Eqs. (7) and (8), written for each of the N nodes of a given finite element, furnish 6N scalar equations, sufficient to permit solution for $\Gamma_1, \ldots, \Gamma_{6N}$ in terms of the 6N nodal deformations. If the nodal numbers of the element are designated $k, i, \ldots, j$ (no sequence implied), and a $(6N \times 1)$ matrix $\bar{y}$ is introduced to represent in the local basis of the element all of the deformational displacements of adjacent nodes, one can construct the matrix equation

$$\bar{y} = F\Gamma \tag{9}$$

with

$$\bar{y} \triangleq \begin{bmatrix} \bar{u}^k \\ \bar{\beta}^k \\ \bar{u}^i \\ \bar{\beta}^i \\ \vdots \\ \bar{u}^j \\ \bar{\beta}^j \end{bmatrix} ; \quad F \triangleq \begin{bmatrix} P|_k \\ \frac{1}{2}\tilde{\nabla}P|_k \\ P|_i \\ \frac{1}{2}\tilde{\nabla}P|_i \\ \vdots \\ P|_j \\ \frac{1}{2}\tilde{\nabla}P|_j \end{bmatrix}$$

where the notation $|_j$ implies evaluation at ξ_j, η_j, ζ_j etc.

Substituting the inverse of Eq. (9) into Eq. (6) yields

$$\bar{w} = PF^{-1}\bar{y} \tag{10}$$

thus establishing the relationship between nodal deformations and the deformations distributed throughout the element. The (3x6N) matrix PF^{-1} which appears frequently in what follows, is called an interpolation matrix and designated W, permitting $\bar{w}$ to be written

$$\bar{w} = W\bar{y} . \tag{11}$$

With full knowledge of the variational deformation field $\bar{w}$ throughout the element, one can obtain an expression for the variational strain field, represented in the local vector basis by $\bar{\varepsilon}_{\alpha\gamma}$ $(\alpha,\gamma = 1,2,3)$. This step requires strain-displacement relationships. When large displacements are considered, as they must be if a steady-state strain due to appendage preload is to be calculated, the nonlinear version of the strain-displacement equations is appropriate. This results in susbstantial analytical complexity, normally circumvented by a process of incremental use of strain-displacement equations linearized about different displacement states. Nonlinearities in the strain-displacement equations are avoided in the present analytical formulation for the solution for small, variational, time-varying deformational displacements by linearizing the strain-displacement equations about the state established by the steady-state preload. Thus the incremental or variational strains in the element beyond any steady-state strains (which will be called $\bar{\varepsilon}'_{\alpha\gamma}$; $\alpha, \gamma = 1,2,3$) can always be related to small variations $\bar{w}_1$, $\bar{w}_2$, $\bar{w}_3$ in displacements with an equation of the form

$$\begin{bmatrix} \bar{\varepsilon}_{11} \\ \bar{\varepsilon}_{22} \\ \bar{\varepsilon}_{33} \\ \bar{\varepsilon}_{12} \\ \bar{\varepsilon}_{23} \\ \bar{\varepsilon}_{13} \end{bmatrix} = D \begin{bmatrix} \bar{w}_1 \\ \bar{w}_2 \\ \bar{w}_3 \end{bmatrix} \quad \text{or} \quad \bar{\varepsilon} = D\bar{w} \tag{12}$$

which becomes

$$\bar{\varepsilon} = DW\bar{y} \tag{13}$$

and when these are small deformational displacements $\bar{w}_1, \bar{w}_2, \bar{w}_3$ corresponding to orthogonal axes ξ, η, ζ Eq. (12) takes the form

$$\begin{bmatrix} \bar{\varepsilon}_{11} \\ \bar{\varepsilon}_{22} \\ \bar{\varepsilon}_{33} \\ \bar{\varepsilon}_{12} \\ \bar{\varepsilon}_{23} \\ \bar{\varepsilon}_{31} \end{bmatrix} = \begin{bmatrix} \partial/\partial\xi & 0 & 0 \\ 0 & \partial/\partial\eta & 0 \\ 0 & 0 & \partial/\partial\zeta \\ \partial/\partial\eta & \partial/\partial\xi & 0 \\ 0 & \partial/\partial\zeta & \partial/\partial\eta \\ \partial/\partial\zeta & 0 & \partial/\partial\xi \end{bmatrix} \begin{bmatrix} \bar{w}_1 \\ \bar{w}_2 \\ \bar{w}_3 \end{bmatrix} . \tag{14}$$

In addition to the variational strain matrix $\bar{\varepsilon}$ above, one may define a steady state strain matrix $\bar{\varepsilon}'$ with six elements chosen from $\bar{\varepsilon}'_{\alpha\gamma}(\alpha, \gamma = 1,2,3)$, and also a strain matrix $\bar{\varepsilon}_\tau$ that would result as a consequence of any deviations from the steady-state thermal condition of the structural appendage. If the deviation from the steady-state temperature at a given point of the element is designated τ , the variational thermal strain $\bar{\varepsilon}_\tau$ becomes

$$\bar{\varepsilon}_\tau = \bar{\alpha}\tau[1 1 1 0 0 0]^T \tag{15}$$

where the scalar $\bar{\alpha}$ is the coefficient of thermal expansion of the element material. When finite element heat transfer equations are introduced to augment the dynamical equations sought here, the distribution of temperature $\tau(\xi, \eta, \zeta)$ in each element would be as sumed to have a simple functional dependence on the nodal temperatures, which become additional unknowns.

The increment $\bar{\sigma}$ in the stress matrix beyond the steady-state value $\bar{\sigma}'$ is related for an elastic material to the difference in the total variational strain and the variational thermal strain by

$$\begin{bmatrix} \bar{\sigma}_{11} \\ \bar{\sigma}_{22} \\ \bar{\sigma}_{33} \\ \bar{\sigma}_{12} \\ \bar{\sigma}_{23} \\ \bar{\sigma}_{31} \end{bmatrix} = \frac{E}{(1+\nu)(1-2\nu)} \begin{bmatrix} (1-\nu) & \nu & \nu & 0 & 0 & 0 \\ \nu & (1-\nu) & \nu & 0 & 0 & 0 \\ \nu & \nu & (1-\nu) & 0 & 0 & 0 \\ 0 & 0 & 0 & (1-2\nu)/2 & 0 & 0 \\ 0 & 0 & 0 & 0 & (1-2\nu)/2 & 0 \\ 0 & 0 & 0 & 0 & 0 & (1-2\nu)/2 \end{bmatrix} \begin{bmatrix} \bar{\varepsilon}_{11} - \bar{\alpha}\tau \\ \bar{\varepsilon}_{22} - \bar{\alpha}\tau \\ \bar{\varepsilon}_{33} - \bar{\alpha}\tau \\ \bar{\varepsilon}_{12} \\ \bar{\varepsilon}_{23} \\ \bar{\varepsilon}_{31} \end{bmatrix} \quad (16)$$

where E is Young's modulus and ν is Poisson's ratio. Symbolically, Eq. (16) may be written

$$\bar{\sigma} = S\bar{\varepsilon} - \bar{\sigma}_{\tau} \quad (17)$$

which with Eq. (13) becomes

$$\bar{\sigma} = SDW\bar{y} - \bar{\sigma}_{\tau} \, . \quad (18)$$

The $(6 \times 6N)$ matrix SDW is sometimes called the element stress matrix.

Variational stresses and strains are related to nodal variational displacements in Eqs. (18) and (13) respective ly. This information can be used in conjunction with the work-energy equation and the virtual displacement concept to obtain expressions for forces and torques that must be applied to the element at the nodes in order to balance the applied loads while sustaining the inertial accelerations associated with nodal accelerations by Eq. (11). Since equal and opposite forces and torques are applied by the elements to the nodal bodies for which equations of motion are to be written in the next section, these expression are the primary immediate objective.

For static equilibrium of a mechanical system the work $\mathscr{W}^*$ accomplished by external forces in the course of a virtual displacement y^* equals the energy U^* stored as strain energy in the deforming element; this equality is preserved for non-dissipative dynamical systems in motion if to the external forces one adds the inertial "force", which for a differential element of volume dv at point p is $-\underline{A}\mu dv$, where $\underline{A}$ is the inertial acceleration of the point p, and μ is the mass density at p. In general, then, the external "forces" doing work include the inertial "forces", the forces and torques applied to the element at its nodes, the body forces (designated by the matrix function $\bar{G}(\xi,\eta,\zeta)$ in the local basis), and the surface forces. In spacecraft applications it is usually sufficient to eliminate the surface loads from participation in $\mathscr{W}^*$ by distributing them to the nodes (as indeed may often be appropriate for the body forces).

For the finite element designated s, let the

$(6N_s \times 1)$ matrix $\bar{L}^s$ be introduced as

$$\bar{L}^s \triangleq \begin{bmatrix} \bar{F}^{ks} \\ \bar{T}^{ks} \\ \\ \bar{F}^{js} \\ \bar{T}^{js} \end{bmatrix} \tag{19}$$

where $\bar{F}^{ks}$ and $\bar{T}^{ks}$ are (3×1) matrices in the local (element) vector basis respectively representing force and torque applied by the kth nodal body to the sth element, and similarly for all N_s nodes of the sth finite element. Thus the work w^* associated with a virtual displacement of the nodes of a generic element becomes

$$w^* = \bar{y}^{*T}\bar{L} + \int \bar{w}^{*T}\bar{G}\,dv - \int \bar{w}^{*T}\bar{A}\mu\,dv$$

where $\bar{A}$ is the (3×1) matrix representing $\underline{A}$ in the local vector basis. With Eq. (11), the work expression becomes

$$w^* = \bar{y}^{*T}\left[\bar{L} + \int W^T(\bar{G} - \bar{A}\mu)\,dv\right]. \tag{20}$$

The incremental strain energy u^* due to the virtual displacement is by virtue of Eqs. (18), (13) and (11) given by

$$U^* = \int \bar{\varepsilon}^{*T}(\bar{\sigma} + \bar{\sigma}')\,dv = \int \bar{w}^{*T}D^T(SDW\bar{y} - \bar{\sigma}_\tau + \bar{\sigma}')\,dv \tag{21}$$

$$= \bar{y}^{*T}\int W^T D^T SDW\,dv\,\bar{y} - \bar{y}^{*T}\int W^T D^T \bar{\sigma}_\tau\,dv + \bar{y}^{*T}\int W^T D^T \bar{\sigma}'\,dv\,.$$

Equating U^* and w^*, dismissing the arbitrary pre-multiplier $\bar{y}^{*T}$ and solving for $\bar{L}$ furnishes

$$\bar{L} = \int W^T D^T S D W d\upsilon \bar{y} + \int W^T [\bar{A}\mu - \bar{G} - D^T \bar{\sigma}_\tau] d\upsilon + \int W^T D^T \bar{\sigma}' d\upsilon . \qquad (22)$$

Note that the last term in Eq. (22) contributes only to the steady-state value of $\bar{L}$.

Eq. (22) is in useful form only when the inertial acceleration matrix $\bar{A}$ is written in terms of the nodal deformation matrix $\bar{y}$ and those functions which define the arbitrary motion of the base b to which the appendage is attached. This is most readily accomplished first in terms of the corresponding Gibbsian vector $\underline{A}$, which by definition is available in terms of the symbols of Fig. 1 as

$$\underline{A} \triangleq \frac{{}^i d^2}{dt}(\underline{X} + \underline{c} + \underline{R} + \underline{R}_c + \underline{\varrho} + \underline{w}) \qquad (23)$$

where the pre-superscript i denotes an inertial reference frame for vector differentiation, and the chain of vectors in parentheses is a single vector locating a differential element of volume in a finite element with respect to an inertially fixed point I. If it should be necessary to identify the particular finite element to which Eq. (23) is being applied, the corresponding numerical superscript can be attached to the vectors $\underline{A}, \underline{R}_c, \underline{\varrho}$ and $\underline{w}$.

Since a matrix formulation is ultimately required, (3×1) matrices are defined for each of the vectors in Eq. (23)

in terms of the most convenient vector basis. In terms of the vector arrays $\{\underline{b}\}$, $\{\underline{a}\}$ and $\{\underline{e}\}$ of Eqs. (2) and (3), and the new array $\{\underline{i}\}$ of inertially fixed unit vectors related to $\{b\}$ by

(24) $$\{\underline{b}\} = \Theta\{\underline{i}\}$$

the vectors in Eq. (23) may be written

$$\begin{aligned} &\underline{X} \triangleq \{\underline{i}\}^T X \qquad &&\underline{R}_c \triangleq \{\underline{a}\}^T R_c \\ &\underline{c} \triangleq \{\underline{b}\}^T c \qquad &&\underline{\rho} \triangleq \{e\}^T \bar{\rho} = \{\underline{a}\}^T \rho \\ &\underline{R} \triangleq \{\underline{a}\}^T R \qquad &&\underline{\omega} \triangleq \{\underline{e}\}^T \bar{w} = \{\underline{a}\}^T w \end{aligned}$$

(25)

thereby defining $X, c, R, R_c, \bar{\rho}, \rho, \bar{w}$ and w.

The inertial reference frame differentiation in Eq. (23) are facilitated by the identity

(26) $$\frac{{}^g d}{dt}\underline{V} = \frac{{}^f d}{dt}\underline{V} + \underline{\omega}^{fg} \times \underline{V}$$

applicable to any vector $\underline{V}$ and any two references frames b and a where $\underline{\omega}^{fg}$ is the angular velocity of f relative to g . With repeated use of Eq. (26), Eq. (23) takes the form

$$\begin{aligned} A = {} & \{\underline{i}\}^T \ddot{X} + \{\underline{b}\}^T \ddot{c} + 2\underline{\omega} \times \{\underline{b}\}^T \dot{c} + \dot{\underline{\omega}} \times \{\underline{b}\}^T c + \underline{\omega} \times (\underline{\omega} \times \{\underline{b}\}^T c) \\ & + \dot{\underline{\omega}} \times \{\underline{b}\}^T R + \underline{\omega} \times (\underline{\omega} \times \{\underline{b}\}^T R) \\ & + \dot{\underline{\omega}}^a \times \{\underline{a}\}^T R_c + \underline{\omega}^a \times (\underline{\omega}^a \times \{\underline{a}\}^T R_c) \\ & + \dot{\underline{\omega}}^a \times \{\underline{a}\}^T \rho + \underline{\omega}^a \times (\underline{\omega}^a \times \{\underline{a}\}^T \rho) \\ & + \{a\}^T \ddot{w} + 2\underline{\omega}^a \times \{\underline{a}\}^T \dot{w} + \dot{\underline{\omega}}^a \times \{a\}^T w + \underline{\omega}^a \times (\underline{\omega}^a \times \{a\}^T w) \ . \end{aligned} \tag{27}$$

where $\underline{\omega}$ and $\underline{\omega}^a$ are the inertial angular velocities of b and a respectively (so that in the more explicit notation of Eq. (26) one would have $\underline{\omega} \triangleq \underline{\omega}^{bi}$ and $\underline{\omega}^a \triangleq \underline{\omega}^{ai}$). Eq. (2) can be used to replace ω and $\underline{\rho}$ in Eq. (27) by $\bar{w}$ and $\bar{\underline{\rho}}$ respectively, and with the introduction of matrices ω and ω^a defined by

$$\underline{\omega} = \{\underline{b}\}^T \omega \,; \qquad \underline{\omega}^a = \{\underline{a}\}^T \omega^a \tag{28}$$

one finds

$$\begin{aligned} \underline{A} = {} & \{\underline{i}\}^T \ddot{X} + \{\underline{b}\}^T [\ddot{c} + 2\tilde{\omega}\dot{c} + (\dot{\tilde{\omega}} + \tilde{\omega}\tilde{\omega})(c + R)] \\ & + \{\underline{a}\}^T (\dot{\tilde{\omega}}^a + \tilde{\omega}^a\tilde{\omega}^a)(R_c + \bar{C}^T \bar{\underline{\rho}}) \\ & + \{\underline{a}\}^T [\bar{C}^T \ddot{\bar{w}} + 2\tilde{\omega}^a \bar{C}^T \dot{\bar{w}} + (\dot{\tilde{\omega}}^a + \tilde{\omega}^a\tilde{\omega}^a)\bar{C}^T \bar{w}] \end{aligned} \tag{29}$$

where tilde on a symbol representing a (3×1) matrix indicates the corresponding (3×3) skew symmetric matrix, e.g.,

$$\tilde{\omega} \triangleq \begin{bmatrix} 0 & -\omega_3 & \omega_2 \\ \omega_3 & 0 & -\omega_1 \\ -\omega_2 & \omega_1 & 0 \end{bmatrix} . \tag{30}$$

Eq. (22) calls for the vector $\underline{A}$ in the vector basis $\{\underline{e}\}$, requiring in Eq. (30) the substitutions from Eqs.(2), (3), and (24)

$$\begin{aligned} \{\underline{a}\}^T &= \{\underline{e}\}^T \bar{C} \\ \{\underline{b}\}^T &= \{\underline{a}\}^T C = \{\underline{e}\}^T \bar{C} C \\ \{\underline{i}\}^T &= \{\underline{b}\}^T \Theta = \{\underline{a}\}^T C\Theta = \{\underline{e}\}^T \bar{C} C \Theta . \end{aligned} \tag{31}$$

From Eqs. (29) and (31) there follows

$$\underline{A} = \{\underline{e}\}^T\bar{A} = \{\underline{e}\}^T\{\bar{C}C\Theta\ddot{X} + \bar{C}C[\ddot{c} + 2\tilde{\omega}\dot{c} + (\dot{\tilde{\omega}} + \tilde{\omega}\tilde{\omega})(c + R)]$$

(32) $$+ \bar{C}(\dot{\tilde{\omega}}^a + \tilde{\omega}^a\tilde{\omega}^a)(R_c + \bar{C}^T\bar{\varrho})$$

$$+ \ddot{w} + \bar{C}[2\tilde{\omega}^a\bar{C}^T\dot{\bar{w}} + (\dot{\tilde{\omega}}^a + \tilde{\omega}^a\tilde{\omega}^a)\bar{C}^T\bar{w}]\} .$$

It should be noted that the quantities $\tilde{\omega}^a, \tilde{\omega}$ and in Eq. (32) are related by the kinematical equations

(33) $$\tilde{\omega}^a = \tilde{\omega} + C\dot{C}^T .$$

Using Eq. (11) to remove $\bar{w}$ from $\bar{A}$, and then substituting for $\bar{A}$ from Eq. (32) into Eq. (22), furnishes

$$\bar{L} = \int W^T D^T SDW dv\,\bar{y} + \int W^T\{\bar{C}C\Theta\ddot{X} + \bar{C}C[\ddot{c} + 2\tilde{\omega}\dot{c} + (\dot{\tilde{\omega}} + \tilde{\omega}\tilde{\omega})(c + R)]$$

(34) $$+ \bar{C}[(\dot{\tilde{\omega}}^a + \tilde{\omega}^a\tilde{\omega}^a)(R_c + \bar{C}^T\bar{\varrho})]\}\mu dv + \int W^T W\mu dv\,\ddot{\bar{y}} + \int W^T\bar{C}2\tilde{\omega}^a\bar{C}^T W\mu dv\,\dot{\bar{y}}$$

$$+ \int W^T\bar{C}(\dot{\tilde{\omega}}^a + \tilde{\omega}^a\tilde{\omega}^a)\bar{C}^T W\mu dv\,\bar{y} - \int W^T(\bar{G} - D^T\bar{\sigma}_\tau)dv + \int W^T D^T\bar{\sigma}' dv .$$

The integrals providing the $(6N \times 6N)$ matrix coefficients of $\ddot{\bar{y}}$, $\dot{\bar{y}}$ and $\bar{y}$ are assigned symbols and labels as follows:

(35) $\bar{m} \triangleq \int W^T W\mu dv$, the element consistent mass matrix

(36) $\bar{g} \triangleq 2\int W^T\bar{C}\tilde{\omega}^a\bar{C}^T W\mu dv$, the element gyroscopic coupling matrix

(37) $\bar{k} \triangleq \int W^T D^T SDW dv$, the element structural stiffness matrix

$\bar{\varkappa} \triangleq \int W^T \bar{C} \tilde{\omega}^a \tilde{\omega}^a \bar{C}^T W \mu \, dv$, the element centripetal stiffness matrix (38)

$\bar{\alpha} \triangleq \int W^T \bar{C} \tilde{\dot{\omega}}^a \bar{C}^T W \mu \, dv$, the element angular acceleration stiffness matrix. (39)

Note that $\bar{m}$, $\bar{k}$ and $\bar{\varkappa}$ are symmetric, while $\bar{g}$ and $\bar{\alpha}$ are skewsymmetric. The bar over these matrices is a reminder that these matrices are associated with the local vector basis $\{\underline{e}\}$. When it becomes necessary to consider these matrices as written for the appendage vector basis $\{\underline{a}\}$, these bars are removed. To obtain m from $\bar{m}$ for example, one may write a transformation written below in terms of the (3×3) submatrices $\bar{C}$ and 0:

$$[m] = \begin{bmatrix} \bar{C}^T & 0 & \dots & 0 \\ 0 & \bar{C}^T & & \vdots \\ \vdots & & \ddots \bar{C}^T & 0 \\ 0 & \dots & 0 & \bar{C}^T \end{bmatrix} [\bar{m}] \begin{bmatrix} \bar{C} & 0 & \dots & 0 \\ 0 & \bar{C} & & \vdots \\ \vdots & & \ddots \bar{C} & 0 \\ 0 & \dots & 0 & \bar{C} \end{bmatrix} . \quad (40)$$

and similarly for $k, \varkappa, g$ and α. The elements of these matrices, such as m_{ij} etc., have indices adopting the $6N$ values associated with the six degrees of freedom of each of the N nodal bodies attached to the element in question.

It may facilitate interpretation to note that the matrices $\bar{C}$ and $\bar{C}^T$ in Eqs. (36)-(39) serve merely to transform the matrix lying between them into the local vector basis.

In application to appendages on a spinning base,

or to otherwise preloaded structures, the matrix $\bar{k}$ is usually considered in the two parts $\bar{k}_0$ and $\bar{k}_\Delta$, with elastic stiffness matrix $\bar{k}_0$ being the stiffness matrix of the element in its unloaded state and with the geometric stiffness matrix or preload stiffness matrix $\bar{k}_\Delta$ accomodating the influence on stiffness attributed to the preload, and often manifested as a consequence of changes in geometry.

Other integrals in Eq. (34) simplify by the removal of terms from the integrand, leaving the matrix $\int W^T \mu dv$. Noting that the deformational displacement of the mass center of the sth element is given in the local vector basis by $\bar{w}_c^s$ in the equation

$$M_s \bar{w}_c^s = \int_s \bar{w} \mu dv = \int_s W \mu dv \bar{y}^s$$

where M_s is the total mass of the sth finite element, one can define the $(3 \times 6N_s)$ matrix W_c^s as the matrix W^s evaluated for the element mass center coordinates $\xi_c^s, \eta_c^s, \zeta_c^s$ and write

$$\int_s W \mu dv = M_s W_c^s . \tag{41}$$

Eq. (34) can now be rewritten in terms of the notation of Eqs. (35)-(41), and now because it will soon become necessary to consider more than one finite element at a time, the superscript s for the sth element will be added where appropriate, furnishing

$$\bar{L}^s = \bar{m}^s \ddot{\bar{y}}^s + \bar{g}^s \dot{\bar{y}}^s + (\bar{k}_0^s + \bar{k}_\Delta^s + \bar{\kappa}^s + \bar{\alpha}^s) \bar{y}^s$$

$$+\int_s W^T\bar{C}(\dot{\tilde{\omega}}^a+\tilde{\omega}^a\tilde{\omega}^a)\bar{C}^T\bar{\varrho}\mu dv+\int_s \bar{W}^T\bar{C}C[\ddot{c}+2\tilde{\omega}\dot{c}+(\dot{\tilde{\omega}}+\tilde{\omega}\tilde{\omega})c]\mu dv$$
$$+M_s W_c^{s^T}\{\bar{C}C\Theta\ddot{X}+\bar{C}^s C(\dot{\tilde{\omega}}+\tilde{\omega}\tilde{\omega})R]+\bar{C}^s(\dot{\tilde{\omega}}^a+\tilde{\omega}^a\tilde{\omega}^a)R_c^s\}$$
$$-\int_s W^T(\bar{G}-D^T\bar{\sigma}_\tau)dv+\int W^T D^T\bar{\sigma}'dv\,. \tag{42}$$

Eq. (42) is still not in the desired final form for $\bar{L}^s$, because the dependence of c on $\bar{y}^s$ has not yet been explicitly accomodated (see Fig. 1 to interpret $-\underline{c}=-\{\underline{b}\}^T$ as the displacement of the vehicle mass center CM from its nominal location in b at point O subsequent to steady-state deformation). The mass center shift $-\underline{c}$ can be attributed in part to the shifts of the mass center locations of the finite elements during deformation, in part to the similar mass center motions of the nodal bodies and in part to the behaviour of moving parts other than the elastic appendage under consideration. If the last of these contributions is symply designated $-\underline{\delta}$, and M represents the total vehicle mass, then by mass center definition

$$\underline{c}=\underline{\delta}-\frac{1}{M}\sum_{i=1}^{n}m_i\underline{u}^i-\frac{1}{M}\sum_{r=1}^{E}M_r\{\underline{e}\}^T\bar{w}_c^r \tag{43}$$

for an appendage with n nodes and E finite elements. Writing both sides of Eq. (43) in the same basis $\{\underline{b}\}$ and substituting from Eq. (40) for $\bar{w}_c^r$ yields

$$\{\underline{b}\}^T c=\{\underline{b}\}^T\left\{\delta-C^T\left[\sum_{i=1}^{n}m_iu^i+\sum_{r=1}^{E}\bar{C}^{r^T}\int W\mu dv\,\bar{y}^r\right]/M\right\} \tag{44}$$

which with Eq. (41) becomes (abandoning the unit vectors)

(45) $$c = \delta - C^T \left[\sum_{i=1}^{n} m_i \underline{u}^i + \sum_{r=1}^{E} \bar{C}^{r^T} M_r W_c^r \bar{y}^r \right] / M \,.$$

Now all terms involving c in Eq. (42) can be removed from the integral over finite element s . Rather than differentiate c as it appears in Eq. (45) to obtain $\dot{c}$ and $\ddot{c}$, one can make further use of Eq. (26), and finally obtain $\bar{L}^s$ from Eq. (42) in the form

(46)
$$\begin{aligned}
\bar{L}^s = {} & \bar{m}^s \ddot{\bar{y}}^s - M_s W_c^{s^T} \bar{C}^{s^T} \left[\sum_{r=1}^{E} \bar{C}^{r^T} M_r W_c^r \ddot{\bar{y}}^r + \sum_{r=1}^{n} m_i \ddot{u} \right] / M \\
& + \bar{g}^s \dot{\bar{y}}^s - 2 M_s W_c^{s^T} \bar{C}^{s^T} \tilde{\omega}^a \left[\sum_{r=1}^{E} \bar{C}^{r^T} M_r W_c^r \dot{\bar{y}}^r + \sum_{i=1}^{n} m_i \dot{u}^i \right] / M \\
& + (\bar{k}_0^s + \bar{k}_\Delta^s + \bar{\varkappa}^s + \bar{\alpha}^s) \bar{y}^s - M_s W_c^{s^T} \bar{C}^{s^T} (\dot{\tilde{\omega}}^a + \tilde{\omega}^a \tilde{\omega}^a) \\
& \cdot \left[\sum_{r=1}^{E} \bar{C}^{r^T} M_r W_c^r \bar{y}^r + \sum_{i=1}^{n} m_i u^i \right] / M \\
& + \int_s W^T \bar{C} (\dot{\tilde{\omega}}^a + \tilde{\omega}^a \tilde{\omega}^a) \bar{C}^T \underline{\varrho} \mu \, dv \\
& + M_s W_c^{s^T} \{ \bar{C}^s C [\Theta \ddot{X} + (\dot{\tilde{\omega}} + \tilde{\omega} \tilde{\omega}) R] + \bar{C}^s (\dot{\tilde{\omega}}^a + \tilde{\omega}^a \tilde{\omega}^a) R_c^s \} \\
& - \int_s W^T (\bar{G} - D^T \bar{\sigma}_\tau) dv + \int_s W^T D^T \bar{\sigma}' dv \\
& + M_c^s W_c^{s^T} \bar{C}^s C [\ddot{\delta} + 2 \tilde{\omega} \dot{\delta} + (\dot{\tilde{\omega}} + \tilde{\omega} \tilde{\omega}) \delta] \,.
\end{aligned}$$

Eq. (46), repeated E times for elements $s = 1,\ldots,E$, provides in the matrices $\bar{L}^1,\ldots,\bar{L}^E$ a representation of the contribution of structural interactions to the forces $\underline{F}^1,\ldots,\underline{F}^n$ and the torques $\underline{T}^1,\ldots,T^n$ applied to the n nodal bodies. There remains the task of deriving equations of motion of these nodal bodies.

b) Nodal body equations of motion

For the jth nodal body, having mass m_j and inertial acceleration $\underline{A}^j$, the translational equation

$$\underline{F}^j = m_j \underline{A}^j \tag{47}$$

can be expressed in the desired form by inspection of the results for a generic point of a finite element. The acceleration $\underline{A}^j$ is defined in terms of the symbols of Fig. 1 as

$$\underline{A}^j \triangleq \frac{{}^i d^2}{dt^2}(\underline{X} + \underline{c} + \underline{R} + \underline{r}^j + \underline{u}^j) \tag{48}$$

which can be compared to Eq. (23) for the element field point. A line of argument parallel to that providing Eq. (29) from Eq. (23) produces from Eq. (48) the expression

$$\begin{aligned} \underline{A}^j = {} & \{\underline{i}\}^T \ddot{X} + \{\underline{b}\}^T [\ddot{c} + 2\tilde{\omega}\dot{c} + (\dot{\tilde{\omega}} + \tilde{\omega}\tilde{\omega})(c + R)] \\ & + \{\underline{a}\}^T [(\dot{\tilde{\omega}}^a + \tilde{\omega}^a\tilde{\omega}^a)(r^j + u^j) + 2\tilde{\omega}^a \dot{u}^j + \ddot{u}^j] \, . \end{aligned} \tag{49}$$

The matrix C can be substituted from Eq. (45), and by the argument leading from there to Eq. (46) one can develop from Eqs. (47) and (49) (with appropriate change of vector basis)

$$
\begin{aligned}
\underline{F}^{j} = \{\underline{a}\}^{T} F^{j} = \{\underline{a}\}^{T} m_{j} \Big\{ & C\Theta\ddot{X} + C[\ddot{\delta} + 2\tilde{\omega}\dot{\delta} + (\dot{\tilde{\omega}} + \tilde{\omega}\tilde{\omega})(\delta + R)] \\
& + (\dot{\tilde{\omega}}^{a} + \tilde{\omega}^{a}\tilde{\omega}^{a}) r^{j} + \ddot{u}^{j} - \Big(\sum_{i=1}^{n} m_{i}\ddot{u}_{i} + \sum_{r=1}^{E} \bar{C}^{r^{T}} M_{r} W_{c}^{r} \ddot{\bar{y}}^{r} \Big) / M \\
& + 2\tilde{\omega}^{a} \Big[\dot{u}^{j} - \Big(\sum_{i=1}^{n} m^{i}\dot{u}^{i} + \sum_{r=1}^{E} \bar{C}^{r^{T}} M_{r} W_{c}^{r} \dot{\bar{y}}^{r} \Big) / M \Big] \\
& + (\dot{\tilde{\omega}}^{a} + \tilde{\omega}^{a}\tilde{\omega}^{a}) \Big[u^{j} - \Big(\sum_{i=1}^{n} m^{i} u^{i} + \sum_{r=1}^{E} \bar{C}^{r^{T}} M_{r} W_{c}^{r} \bar{y}^{r} \Big) / M \Big] \Big\} .
\end{aligned} \tag{50}
$$

The force $\underline{F}^{j}$ applied to the jth nodal body consists of the external force $\underline{f}^{j} = \{\underline{a}\}^{T} f^{j}$ applied at that node plus the structural interaction forces $\underline{F}^{sj}$ applied to node j by adjacent structural elements s . If the symbol $\sum_{s\varepsilon E_j}$ denotes summation over those values of s belonging to the set E_j consisting of that subset of the element numbers $1,\ldots,E$ corresponding to elements in contact with node j , then $\underline{F}^{j}$ becomes

$$
\underline{F}^{j} = \underline{f}^{j} + \sum_{s\varepsilon E_j} \underline{F}^{sj} . \tag{51}
$$

If $\underline{F}^{sj}$ is written in the vector basis $\{\underline{e}^{s}\}$ as

$$
\underline{F}^{sj} \triangleq \{\underline{e}^{s}\}^{T} \bar{F}^{sj} = \{\underline{a}\}^{T} \bar{C}^{s^{T}} \bar{F}^{sj} \tag{52}
$$

and the relationship $\bar{F}^{sj} = -\bar{F}^{js}$ is accepted as a consequence of Newton's third law, one can extract from Eq. (50) the matrix e-

quations

$$f^j - \sum_{s \in E_j} \bar{C}^{s^T} \bar{F}^{js} = m^j \Big\{ C\Theta\ddot{X} + C[\ddot{\delta} + 2\tilde{\omega}\dot{\delta} + (\dot{\tilde{\omega}} + \tilde{\omega}\tilde{\omega})(\delta + R)]$$

$$+ (\dot{\tilde{\omega}}^a + \tilde{\omega}^a\tilde{\omega}^a)r^j + \ddot{u}^j - \Big(\sum_{i=1}^{n} m^i\ddot{u}^i + \sum_{r=1}^{E} \bar{C}^{r^T} M_r W_c^r \ddot{\bar{y}}^r\Big)/M$$

$$+ 2\tilde{\omega}^a\Big[\dot{u}^j - \Big(\sum_{i=1}^{n} m^i\dot{u}^i + \sum_{r=1}^{E} \bar{C}^{r^T} M_r W_c^r \dot{\bar{y}}^r\Big)/M\Big]$$

$$+ (\dot{\tilde{\omega}}^a + \tilde{\omega}^a\tilde{\omega}^a)\Big[u^j - \Big(\sum_{i=1}^{n} m^i u^i + \sum_{r=1}^{E} \bar{C}^{r^T} M_r W_c^r\Big)/M\Big]\Big\}$$

$$j = 1, \ldots, n \tag{53}$$

here for convenience in future composition of matrix equations the (3×3) unit matrix U has been used to define the mass matrix

$$m^j = m_j U . \tag{54}$$

By systematically examining the quantities $\bar{L}$ defined in Eq. (19) and appearing in the E matrix equations represented by Eq. (46), one can extract expressions for the quantities $\bar{F}^{js}$ appearing in Eq. (53); upon substitution of these expressions one has in Eq. (53) a set of dynamical equations in u^j and $\bar{y}^s$, $j=1,\ldots,n$, $s=1,\ldots,E$. By the definition found after Eq. (9), the matrices $\bar{y}^1,\ldots,\bar{y}^E$ are comprised of the matrices $\bar{u}^1,\ldots,\bar{u}^n,\bar{\beta}^1,\ldots,\bar{\beta}^n$ which transform to u^i and β^i by $\bar{u}^i = \bar{C}u^i$ and $\bar{\beta}^i = \bar{C}\beta^i$, $i=1,\ldots,n$. Thus Eq. (53), with substitutions from Eq. (46), provides $3n$ scal

ar second order differential equations in the 6N unknowns $u_\alpha^1, \ldots, u_\alpha^n, \beta_\alpha^1, \ldots, \beta_\alpha^n$ $(\alpha = 1,2,3)$. Completion of the set requires the equations of rotational motion of the nodal bodies.

The basic equation for the rotation of the jth nodal rigid body is

$$\underline{T}^j = \dot{\underline{H}}^j = \square^j \cdot \dot{\underline{\omega}}^j + \dot{\square}^j \cdot \underline{\omega}^j = \square^j \cdot \dot{\underline{\omega}}^j + \underline{\omega}^j \times \square^j \cdot \underline{\omega}^j \tag{55}$$

where $\underline{T}^j$ is the applied torque, $\underline{H}^j$ the angular momentum, and $\square^j$ the inertia dyadic of the nodal body, all referred to the mass center of the body, and over-dot denotes time differentiation in an inertial frame of reference. The inertial angular velocity ω^j of the jth body may be expressed in terms of established notation as

$$\underline{\omega}^j = \underline{\omega}^a + \{\underline{a}\}^T \dot{\beta}^j = \{\underline{a}\}^T (\omega^a + \dot{\beta}^j) \tag{56}$$

and its inertial derivative is

$$\dot{\underline{\omega}}^j = \dot{\underline{\omega}}^a + \{\dot{\underline{a}}\}^T \ddot{\beta}^j + \underline{\omega}^a \times \{\underline{a}\}^T \dot{\beta}^j = \{\underline{a}\}^T (\dot{\omega}^a + \ddot{\beta}^j + \tilde{\omega}^a \dot{\beta}^j) \tag{57}$$

so that Eq. (55) becomes

$$\begin{aligned} \underline{T}^j = \{\underline{a}\}^T T^j &= \{\underline{n}^j\}^T I^j \{\underline{n}^j\} \cdot \{\underline{a}\}^T (\dot{\omega}^a + \ddot{\beta}^j + \tilde{\omega}^a \dot{\beta}^j) \\ &+ \{\underline{a}\}^T (\omega^a + \dot{\beta}^j) \times \{\underline{n}^j\}^T I^j \{\underline{n}^j\} \cdot \{\underline{a}\}^T (\omega^a + \dot{\beta}^j) \end{aligned} \tag{58}$$

where $\{\underline{n}^j\}$ is the (3×1) array of dextral, orthogonal, unit vectors $\underline{n}_1^j$, $\underline{n}_2^j$, $\underline{n}_3^j$ fixed in nodal body j, and coincident with $\{\underline{a}\}$

when the appendage is in its steady-state (see Fig. 1). The direction cosine matrix relating $\{\underline{n}^j\}$ and $\{\underline{a}\}$ subsequent to small appendage deformation is given by the relationship

$$\{\underline{n}^j\} = (U - \tilde{\beta}^j)\{\underline{a}\} \tag{59}$$

where U is the (3×3) unit matrix and $\tilde{\beta}^j$ is the skew-symmetric matrix formed of the elements $\beta_1^j, \beta_2^j, \beta_3^j$ according to the pattern of Eq. (30), i.e. $\tilde{\beta}^j_{\alpha\theta} \triangleq \varepsilon_{\alpha\gamma\theta}\beta^j_\gamma$ where $\varepsilon_{\alpha\gamma\theta}$ is the epsilon symbol of tensor analysis.

Substituting Eq. (59) into Eq. (58) produces a vector equation entirely in the $\{\underline{a}\}$ basis, or equivalently the matrix equation

$$\begin{aligned} T^j = {} & I^j\dot{\omega}^a + \tilde{\omega}^a I^j \omega^a + I^j\ddot{\beta} + [\tilde{\omega}^a I^j - (I^j\omega^a)^\sim + I^j\tilde{\omega}^a]\dot{\beta}^j \\ & + [I^j\dot{\omega}^a - (I^j\dot{\omega}^a)^\sim + \tilde{\omega}^a I^j \tilde{\omega}^a - \tilde{\omega}^a(I^j\omega^a)^\sim]\beta^j \end{aligned} \tag{60}$$

where second degree terms in the matrix β^j and its derivatives have been ignored, and the tilde retains its operational significance (see Eq. (30)), so that for example $(I^j\omega^a)^\sim_{\alpha\theta} \triangleq \varepsilon_{\alpha\gamma\theta} I^j_{\gamma\eta}\omega^a_\eta$.

The torque $\underline{T}^j$ applied to the jth nodal body consits of the external torque $\underline{t}^j = \{\underline{a}\}^T t^j$ applied at that node plus the structural interaction torques $\underline{T}^{sj}$ applied to node j by adjacent structural elements s . If as in Eq. (51) the set E_j contains the numbers of the elements in contact with node j , then $\underline{T}^j$ may be written (in parallel with Eqs. (51), (52)) as

$$\underline{T}^{j} = \{\underline{a}\}^T T^{j} = \{\underline{a}\}^T t^{j} + \sum_{s\varepsilon E_j} \underline{T}^{sj} = \{\underline{a}\}^T t^{j} - \sum_{s\varepsilon E_j} \underline{T}^{js}$$

$$(61) \qquad = \{\underline{a}\}^T \left[t^{j} - \sum_{s\varepsilon E_j} \bar{C}^{s^T} \bar{T}^{js} \right] .$$

The combination of Eqs. (60) and (61) provides

$$t^{j} - \sum_{s\varepsilon E_j} \bar{C}^{s^T} \bar{T}^{js} = I^{j}\dot{\omega}^a + \tilde{\omega}^a I^{j}\omega^a + I^{j}\ddot{\beta}^{j} + [\tilde{\omega}^a I^{j} - (I^{j}\omega^a)^{\sim} + I^{j}\tilde{\omega}^a]\dot{\beta}^{j}$$

$$(62) \qquad + [I^{j}\dot{\tilde{\omega}}^a - (I^{j}\dot{\omega}^a)^{\sim} + \tilde{\omega}^a I^{j}\tilde{\omega}^a - \tilde{\omega}^a(I^{j}\omega^a)^{\sim}]\beta^{j}$$

$$j = 1,\ldots,n .$$

The rotational Eqs. (62) stand in parallel with the transational Eqs. (53) as the basic equations of motion of the n nodal bodies of the appendage. Once Eqs. (46) and (19) have been used to provide expressions for the matrices $\bar{T}^{js}$ and $\bar{F}^{js}$ appearing respectively in Eqs. (62) and (53), these constitute a complete set of dynamical equations.

4. Coordinate transformations

There remains the critical task of packaging Eqs. (53) and (62), with substitutions from Eq. (46), in a form convenient for the generation of coordinate transformations. To this end, let

$$q \triangleq [u_1^1 u_2^1 u_3^1 \beta_1^1 \beta_2^1 \beta_3^1 u_1^2 \dots \beta_3^n]^T \tag{63}$$

be the $(6n \times 1)$ matrix of nodal deformation coordinates, and rewrite the $6n$ second order differential equations implied by Eqs. (46), (53), and (62) in the form

$$M'\ddot{q} + D'\dot{q} + G'\dot{q} + K'q + A'q = L' \tag{64}$$

where M', D' and K' are $(6n \times 6n)$ symmetric matrices and where G' and A' are $(6n \times 6n)$ skew-symmetric matrices, with L' a $(6n \times 1)$ matrix not involving the deformation variables in q. Since Eqs. (53), (62), and (46) are all linear in the variables u^j, β^j and $\bar{y}^j$ contained within q, and since any square matrix can be written as the sum of symmetric and skew-symmetric parts, the possibility of expression of these equations in the form of Eq. (64) is guaranteed by the symmetric character of the coefficients of $\ddot{u}^j, \ddot{\beta}^j$ and $\ddot{\bar{y}}^j$ in the constituent equations.

The $(6n \times 6n)$ matrix M' can be represented as the sum of three parts, as symbolized by

$$M' \triangleq M + M^c - \bar{M} \tag{65}$$

where M is null except for the (3×3) matrices $m^1, I^1, m^2, \dots, I^n$ along its principal diagonal, M^c is the consistent mass matrix whose elements M_{ij}^c are given in terms of the constituents of the finite element inertia matrices m in Eq. (40) by

(66) $$M^c_{ij} = \sum_{s=1}^{E} m^s_{ij}$$

and the contribution $-\bar{M}$ accomodates the reduction of the effective inertia matrix due to mass center shifts within the vehicle induced by deformation (see for example the terms

$$-\left(\sum_{i=1}^{n} m^i \ddot{u}^i + \sum_{r=\ell}^{E} \bar{C}^{r^T} M_r W^r_c \ddot{\bar{y}}^r\right)/M$$ in Eq. (53)).

The matrix D' in Eq. (64) accomodates any viscous damping that may be introduced to represent energy dissipation due to structural vibrations. As the equations (62), (53), and (46) have been formulated here, such terms have been omitted, but they can still be inserted if one accepts the practice common among structural dynamicists of incorporating the equivalent of a term $D'\dot{q}$ into equations of vibration only after derivation of equations of motion and transformation of coordinates.

Examination of the coefficients of $\dot{\beta}^j$, $\dot{u}^j$ and $\dot{\bar{y}}^j$ in Eqs. (62), (53), and (46) reveals that all have coefficients which will appear in the skew-symmetric matrix G' in Eq. (64) (*); since all such terms disppear when ω^a is nominally zero, the matrix G' is said to provide the gyroscopic coupling of the equa-

(*) The identity $\tilde{\omega}^a I^j - (I^j \omega^a)^\sim + I^j \tilde{\omega}^a = (\mathrm{tr}\, I^j)\tilde{\omega}^a - 2(I^j\omega^a)^\sim$ is required in Eq. (62) to reveal the skew symmetry of the coefficient of $\dot{\beta}^j$.

tions of vibration. Note that the matrices $\bar{g}^s$ defined generically in Eq. (37) contribute to G' just as the matrices $\bar{m}^s$ contribute to M' (see Eqs. (65), (66)).

The terms from Eqs. (62), (53), and (46) contributing to the matrix K' in Eq. (64) are basically of three kinds: i) those represented by $\bar{k}_0^s$ in Eq. (46), which reflect the elastic stiffness of the structure in its unloaded state; ii) those represented by $\bar{k}_\Delta^s$ in Eq. (46), which provide the increment to the stiffness of the structure attributable to structural preload; and iii) those represented in Eq. (46) by $\bar{\kappa}^s$ and in Eqs. (46), (53), and (62) by other terms involving base acceleration (such as the centripetal acceleration term $m^{\jmath}\tilde{\omega}^a\tilde{\omega}^a$ in Eq. (53)). The elements of the matrices $\bar{k}_0^s$, $\bar{k}_\Delta^s$, and $\bar{\kappa}^s$ contribute to K' in a manner analogous to the contribution of $\bar{m}^s$ to M' (see Eqs. (65), (66)).

Finally, the matrix A' in Eq. (64) contains all terms from Eqs. (46), (53), and (62) involving $\dot{\omega}^a$, and in addition the coefficient $-\tilde{\omega}^a(I^{\jmath}\omega^a)^\sim$ of $\dot{\beta}^{\jmath}$ in Eq. (62) makes a contribution to A'. Because certain of the coordinate transformation procedures to be considered depend upon the absence of the matrix A', it is worthwile to examine the skew-symmetric part of the matrix $-\tilde{\omega}^a(I^{\jmath}\omega^a)^\sim$ in detail, since when ω^a has some nominal constant value, say Ω, and $\dot{\omega}^a$ is nominally zero, this matrix is the sole contributor to A'. In terms of its symmetric and skew-symmetric parts, this matrix is

$$-\tilde{\omega}^a(I^\phi\omega^a)^\sim = -\frac{1}{2}\left[\tilde{\omega}^a(I^\phi\omega^a)^\sim + (I^\phi\omega^a)^\sim\tilde{\omega}^a\right]$$

$$-\frac{1}{2}\left[\tilde{\omega}^a(I^\phi\omega^a)^\sim - (I^\phi\omega^a)^\sim\tilde{\omega}^a\right] . \tag{67}$$

The matrix identity

$$\tilde{x}\tilde{y} - \tilde{y}\tilde{x} = (\tilde{x}y)^\sim \tag{68}$$

for any (3×1) matrices x and y permits the skew-symmetric part of $-\tilde{\omega}^a(I^\phi\omega^a)^\sim$ to be recorded as

$$-\frac{1}{2}[\tilde{\omega}^a(I^\phi\omega^a)^\sim - (I^\phi\omega^a)^\sim\tilde{\omega}^a] = -\frac{1}{2}[\tilde{\omega}^a I^\phi\omega^a]^\sim \approx -\frac{1}{2}[\tilde{\Omega}I^\phi\Omega]^\sim \tag{69}$$

where the final substitution replaces ω^a by its nominal value Ω. In terms of scalars representing the elements $I_{\alpha\theta}$ of I^ϕ and Ω_θ of $\bar{\Omega}$, $\alpha(\theta = 1,2,3)$ the independent nonzero terms of $-\frac{1}{2}[\tilde{\Omega}I^\phi\Omega]^\sim$ are given by

$$-\frac{1}{2}[\tilde{\Omega}I^\phi\Omega]^\sim_{12} = -\frac{1}{2}[(I_{11}-I_{22})\Omega_1\Omega_2+I_{12}(\Omega_2^2-\Omega_1^2)+I_{13}\Omega_2\Omega_3-I_{23}\Omega_1\Omega_3]$$

$$-\frac{1}{2}[\tilde{\Omega}I^\phi\Omega]^\sim_{13} = -\frac{1}{2}[(I_{11}-I_{33})\Omega_1\Omega_3+I_{13}(\Omega_3^2-\Omega_1^2)+I_{12}\Omega_2\Omega_3-I_{32}\Omega_1\Omega_2] \tag{70}$$

and

$$-\frac{1}{2}[\tilde{\Omega}I^\phi\Omega]^\sim_{23} = -\frac{1}{2}[(I_{22}-I_{33})\Omega_2\Omega_3+I_{23}(\Omega_3^2-\Omega_2^2)+I_{21}\Omega_1\Omega_3+I_{31}\Omega_3\Omega_2].$$

Since such terms as these are the sole contributors to A' when is nominally zero, it becomes clear that the special case $A' = 0$ applies when the base experiences small excursions about a non-zero constant spin only if the nodal bodies are particles or spheres (or in the extraordinary case when in the steady-state of deformation all nodal bodies have principal axes of inertia aligned with the nominal value of the angular velocity $\underline{\omega}^a$).

The objective of this section is to find a coordinate transformation which will permit the replacement of the homogeneous form of Eq. (64) with a set of completely uncoupled differential equations. Although the conceptual, analytical, and computational difficulties encountered in meeting this objective in general terms are greatly diminished in special cases of practical interest, consideration will be given here only to the most general tractable case of Eq. (64) and to a special case of Eq. (64) for which $A' = D' = 0$.

Inspection of Eqs. (62), (53), and (46) reveals that the coefficients of q and $\dot{q}$ in Eq. (64) depend upon ω^a , which characterizes the rotational motion of the appendage base. For the problems of interest, ω^a is an unknown fucntion of time, to be determinated only after the appendage Eqs. (64) are augmented by other equations of dynamics and control for the total vehicle and solved. Only if ω^a can be assumed to experience, in a given time interval, small excursions about a constant nominal value (say Ω) is there any possibility of obtaining from Eq. (64)

a transformation to uncoupled equations. Any methods involving modal coordinates (see Introduction) are dependent upon this assumption, adopted henceforth. With this restriction, the coefficient matrices of q, $\dot{q}$ and $\ddot{q}$ in Eq. (64) are constants, since products of small quantities are to be ignored.

If all of the matrices A', K', G', D' and M' in Eq. (64) are constant but nonzero, there exists no transformation of the form $q = \Phi\eta$ with η a $(6n \times 1)$ matrix of new coordinates, which can be used to obtain from Eq. (64) a second order differential equation in η with diagonal coefficient matrices. In order to transform Eq. (64) to a set of uncoupled equations it is first necessary to rewrite Eq. (64) in first order form, such as

$$A\dot{Q} + BQ = L \tag{71}$$

where

$$Q \triangleq \begin{bmatrix} q \\ \hline \dot{q} \end{bmatrix}; \quad L \triangleq \begin{bmatrix} 0 \\ \hline L' \end{bmatrix}$$

$$A \triangleq \left[\begin{array}{c|c} K' + A' & 0 \\ \hline 0 & M' \end{array}\right]; \quad B \triangleq \left[\begin{array}{c|c} 0 & -K' - A' \\ \hline K' + A' & D' + G' \end{array}\right].$$

Now let Φ be a $(12n \times 12n)$ matrix of (complex) eigenvectors of the differential operator in Eq. (71), and let Φ' be a $(12n \times 12n)$

matrix of (complex) eigenvectors of the homogeneous adjoint equation

$$A^T \dot{Q}' + B^T Q' = 0 \tag{72}$$

so that Φ and Φ' are related by

$$\Phi^{-1} = \ell \Phi'^T \tag{73}$$

with ℓ a $(12n \times 12n)$ diagonal matrix which depends upon the normalization of Φ and Φ'. Substitution into Eq. (64) of the transformation

$$Q = \Phi Y \tag{74}$$

and pre-multiplication by Φ'^T furnishes

$$(\Phi'^T A \Phi)\dot{Y} + (\Phi'^T B \Phi)Y = \Phi'^T L \, . \tag{75}$$

The two coefficient matrices enclosed in parentheses are diagonal (as is evident from Eq. (73) when $A=U$, which by virtue of the nonsingularity of A can be assumed for this proof without loss of generality.) If Λ is the $(12n \times 12n)$ matrix of the (complex) eigenvalues of the differential operator in Eq. (71) (or Eq. (72), which has the same eigenvalues), then upon pre-multiplication by $(\Phi'^T A \Phi)^{-1}$ one obtains

$$\dot{Y} = \Lambda Y + (\Phi'^T A \Phi)^{-1} \Phi'^T L \tag{76}$$

which is in a form convenient for computation. (Note that the matrix inversion in Eq. (76) consists simply of calculating the reciprocals of the diagonal elements of $\Phi'^T A \Phi$.) In practice, one may expect that physical interpretation of the new (complex) state variables $Y_1,\dots,Y_{12n}$ (see Ref. 7) will permit truncation to a reduced set of variables contained in a new $(2n \times 1)$ matrix $\bar{Y}$, and with corresponding truncation of Λ to the $(2n \times 2n)$ matrix $\bar{\Lambda}$ and truncation of Φ and Φ' to the $(12N \times 2N)$ matrices $\bar{\Phi}$ and $\bar{\Phi}'$, one can reduce Eq. (76) to

$$\dot{\bar{Y}} = \overline{\Lambda Y} + (\bar{\Phi}'^T A \bar{\Phi})^{-1} \bar{\Phi}'^T L . \tag{77}$$

Eq. (77) may be used in conjunction with vehicle equations of motion to simulate system behavior.

In the special case for which $A' = D' = 0$, the matrices A and B in Eq. (71) are respectively symmetric and skew symmetric, so that Eq. (72) behaviour.

$$A\dot{Q}' - BQ' = 0 \tag{78}$$

and the adjoint eigenvector matrix Φ is available immediately as the complex conjugate (*) of Φ . After truncation this result can be substituted into Eq. (77), so that in this special case the final equations are obtained without the requirement of actually computing the eigenvectors in Φ' . Although transformations other than Eq. (74) can also be applied in this special

(*) This observation is a contribution of Mr. A.S. Hopkins of UCLA and McDonnel-Douglas Corp.

case with $A' = D' = 0$, the advantage would appear to be with Eq. (74). Transformations superior to Eq. (74) are well known to exist when $A' = G' = 0$, and particularly so when D' is a polynomial in M' and K'; in this last and simplest case, one can accomplish a transformation to dynamically uncoupled coordinates even at the level of second order differential equations.

Specifically, if an elastic appendage is vibrating about a state of rest in inertial space, Eq. (64) becomes

$$M'\ddot{q} + K'q = L' . \tag{79}$$

If the eigenvector of Eq. (79) are assembled as the columns of the matrix Φ , then the transformation

$$q = \Phi\eta \tag{80}$$

followed by premultiplication by Φ^T provides (with suitable normalization of the eigenvectors in

$$\ddot{\eta} + \sigma^2\eta = \Phi^T L' \tag{81}$$

where σ^2 is a diagonal matrix whose nonzero elements are the negative squares of the eigenvalues of Eq. (79), so σ_j is the jth "natural frequency" of the structure.

Since the scalar equations implied by Equation (81) are uncoupled, it is quite reasonable to consider them independently and to truncate the coordinate matrix η to a smaller matrix $\bar{\eta}$, retaining only those distributed (or modal) coordinates $\eta_1,\dots,\eta_N$ which seem required for dynamic simulation. At this point it is customary to introduce modal damping by means of the diagonal matrix $\bar{\zeta}$, to obtain

$$\ddot{\bar{\eta}} + 2\bar{\zeta}\bar{\sigma}\dot{\bar{\eta}} + \bar{\sigma}^2\bar{\eta} = \bar{\Phi}^T L' . \tag{82}$$

The implication of the insertion of the damping term $2\bar{\zeta}\bar{\sigma}\dot{\bar{\eta}}$ is that Eq. (79) has been augmented by a term $D'\dot{q}$, with D' a polynomial in M' and K' (usually considered as a linear combination of M' and K', so $D'\dot{q}$ is called a proportional damping term).

5. The influence of spin on mode shapes and frequencies

A qualitative appreciation of the influence of spin on structural oscillations can be gained by consideration of the simple systems illustrated in Fig. 2. In each of the figures 2a), 2b), 2c) the dynamical system is a single particle P of mass m, constrainted to move in a massless tube and elastically connected to a rigid structure b which has a prescribed constant inertial angular velocity $\Omega \underline{b}_3$ (unit vectors b_1, b_2, b_3 are fixed in b). Springs with effective translational spring constants k_1, k_2, k_3 in each case resist the displacement of P through small displacement vector $\underline{u} = u_1\underline{b}_1 + u_2\underline{b}_2 + u_3\underline{b}_3$ from its nominal or steady-state location P' relative to b. The spring identified by k_2 is deformed even in the steady-state location due to the

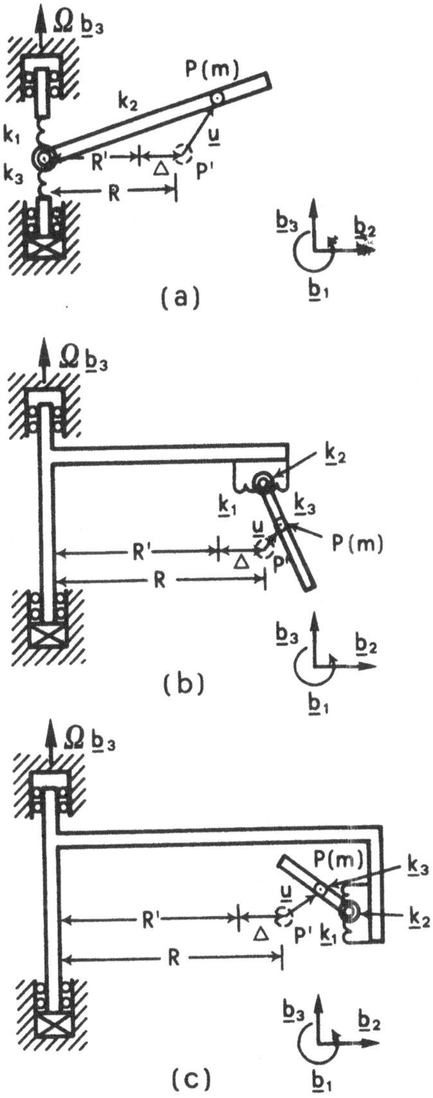
Ω b3
P(m)
k2
k1
u
R'
Δ
P'
k3
R
b3
b2
b1
(a)
Ω b3
k2
k1
k3
u
P(m)
R'
Δ
P'
R
b3
b2
b1
(b)
Ω b3
P(m)
k3
u
R'
Δ
P'
k1
k2
R
b3
b2
b1
(c)

Fig. 2

centrifugal "force" induced by spin; if Ω were zero the location of P' would be a small distance Δ closer to the bearing axis of b, where $k_2\Delta = mR\Omega^2$. The only significant difference in systems 2a), 2b), and 2c) is in the steady-state load in the spring restraining the motion of P within the tube; in case 2a) this spring is in tension, in case 2b) it is unloaded, and in case 2c) it is in compression.

For each of the three cases a), b), c), linearized equations of motion in the small deformation variables u_1, u_2 u_3 are

$$m\ddot{u}_1 - 2m\Omega\dot{u}_2 + (k_1' - m\Omega^2)u_1 = 0 \tag{83}$$

$$m\ddot{u}_2 + 2m\Omega\dot{u}_1 + (k_2' - m\Omega^2)u_2 = 0 \tag{84}$$

$$m\ddot{u}_3 + k_3' u_3 = 0 \tag{85}$$

where k_1', k_2', k_3' are effective values of k_1, k_2, k_3 modified to accomodate the influence (if any) of structural preload due to centrifugal "forces" induced by spin. These modified spring constants are different for cases 2a), 2b), and 2c).

Specifically, in case a) there is a structural preload in the spring within the tube of magnitude $k_2\Delta = mR\Omega^2$; in response to displacement u_3 this force acquires a restoring component of magnitude $m\Omega^2 u_3$ and in response to a displacement u_1 a restoring force of magnitude $m\Omega^2 u_1$ develops. Thus in case a) we have $k_1' = k_1 + m\Omega^2$, $k_2' = k_2$ and $k_3' = k_3 + m\Omega^2$. With the definitions $\omega_1^2 \triangleq k_1/m$, $\Delta_2^2 \triangleq k_2/m$, $\omega_3^2 \triangleq k_3/m$, Eqs. (83) - (85) become for the system of Figure 2a)

$$\ddot{u}_1 - 2\Omega\dot{u}_2 + \omega_1^2 u_1 = 0 \tag{86}$$

$$\ddot{u}_2 + 2\Omega\dot{u}_1 + (\omega_1^2 - \Omega^2)u_2 = 0 \tag{87}$$

$$\ddot{u}_3 + (\omega_3^2 + \Omega^2)u_3 = 0 . \tag{88}$$

In case b), on the other hand, the preload in the spring identified by k_2 has no influence on the restoring forces developed after displacement u_1 , u_2 or u_3; hence the equations of motion are

$$\ddot{u}_1 - 2\Omega\dot{u}_2 + (\omega_1^2 - \Omega^2)u_1 = 0 \tag{89}$$

$$\ddot{u}_2 + 2\Omega\dot{u}_1 + (\omega_1^2 - \Omega^2)u_2 = 0 \tag{90}$$

$$\ddot{u}_3 + \omega_3^2 u_3 = 0 . \tag{91}$$

In case c) the preload in the k_2 spring acts to <u>increase</u> displacements u_1 and u_3 , giving rise to equations of motion

$$\ddot{u}_1 - 2\Omega\dot{u}_2 + (\omega_1^2 - 2\Omega^2)u_1 = 0 \tag{92}$$

$$\ddot{u}_2 + 2\Omega\dot{u}_1 + (\omega_2^2 - \Omega^2)u_2 = 0 \tag{93}$$

$$\ddot{u}_3 + (\omega_3^2 - \Omega^2)u_3 = 0 . \tag{94}$$

Thus we can see that the influence of spin-induced preloads is dependent upon the structural configuration, acting sometimes to stiffen the structure (and elevate its natural frequency), and sometimes to soften it (reducing the natural frequency of vibration). In generalizing from this simple example to an arbitrary structure, we find the influence of preload manifested as a structural preload stiffness matrix (more often called the geometric stiffness matrix), as discussed in the preceding text following Eq. (40).

The influence of centripetal acceleration is represented by Eq. (38) in the general case, and by the terms $-m\Omega^2 u_1$ and $-m\Omega^2 u_2$ in Eqs. (83) and (84) for the systems of Figure 2. Note that these terms effectively reduce stiffness in directions normal to the spin axis.

Coriolis accelerations provide the terms identified in Eq. (36) in the general case, and the term $-2m\Omega\dot{u}_2$ and $2m\Omega\dot{u}_1$ in the special cases of Figure 2. These terms couple oscillations in the plane normal to the spin axis, providing complex eigenvectors denoting vibratory modes with phase lags. It is easily shown by examination of the characteristic equation that the influence of the Coriolis terms on the natural frequencies of vibration in the plane normal to the spin axis is to elevate the higher of these two frequencies and to diminish the lower of the two.

6. Equations of motion collections of substructures

As noted in Section 2.1, modern spacecraft are often idealized as collections of substructures, each of which might be idealized in a wide variety of ways. In Section 2.3 on substructure equations, it was assumed in order to focus that the substructure was a finite element system and the adjacent body was rigid, but this is a very special case. The procedure to be followed in developing equations for systems of substructures depends very much on the substructure models. Various approaches will be outlined here without detailed development.

If the vehicle is idealized as a rigid body with one or more flexible appendage, with gross relative motions kinematically prescribed (so $C(t)$ in Eq. (2) is prescribed), then the substructure equations of each appendage can be augmented by applying $\underline{F} = m\underline{A}$ and $\underline{T} = \dot{\underline{H}}$ to the <u>total</u> <u>vehicle</u>, so that $\underline{F}$ is the resultant external force and $\underline{T}$ the resultant external torque for the vehicle mass center, A is the mass center inertial acceleration, m is the vehicle mass and $\underline{H}$ the system angular momentum about the vehicle mass center.

If the vehicle is idealized as a rigid body with one or more attached flexible appendages being subject to gross relative motions caused by control system torques, then one must apply $\underline{F} = m\underline{A}$ and $\underline{T} = \underline{H}$ not to the total vehicle but vehicle with automatically controlled appendages removed. This procedure requires that special precautions be taken to preserve the integrity of appendage forces and torques after coordinate truncation.

If the vehicle is idealized as a collection of elastic substructures, it becomes necessary to introduce constraint equations to maintain compatibility of deformations at the interfaces between substructures. Although much remains to be done in the development of analytical procedures for this case, any approach must involve firstly the introduction of "constraint modes" in order to represent the consequences of deformations violating the boundary conditions assumed for modal analysis, and secondly the use of constraint equations in the reduction of coordinates involved redundantly in the constraint modes of connected elastic substructures.

7. Dynamic analysis for control design

Although any system of continuous differential equations with a known solution can be written as equations linearized in the variational coordinates for that solution, and these equations can always be used to provide matrix transfer functions among the variables, this approach is particularly fruitful when scalar transfer functions are extracted to relate the vehicle rotation about a given axis. Such transfer functions may be expected to provide results indicative of spacecraft dynamic response when the vehicle is nonspinning and the dominant modes of appendage vibrations do not produce excessive cross-axis coupling of the spacecraft.

If $\underline{\theta} \triangleq [\theta_1\ \theta_2\ \theta_3]^T$ is the matrix of 1-2-3 attitude angles of a rigid body with flexible appendages, and $\underline{I}$ is the in ertia matrix of the total vehicle for the vehicle mass center then the vehicle response to external torque $\underline{T}$ about the mass center is established by the linearized equation

$$T = I\ddot{\theta} - \bar{\delta}^T \ddot{\bar{\eta}} \qquad (95)$$

and the appendage modal coordinates in the matrix $\bar{\eta}$ as, from Eq. (82),

$$\ddot{\bar{\eta}} + 2\bar{\varrho}\bar{\sigma}\dot{\bar{\eta}} + \bar{\sigma}^2\bar{\eta} = \bar{\delta}\ddot{\theta} \qquad (96)$$

where

$$\bar{\delta} \triangleq -\phi^T M(\Sigma_{0U} - \Sigma_{U0}\tilde{R} - \tilde{r}\Sigma_{U0}) \qquad (97)$$

with M as defined in Eq. (65), R as defined in Eq. (25), $\tilde{r}$ a $6n \times 6n$ matrix null except for the 3×3 matrices $\tilde{r}^1, 0, \tilde{r}^2, 0 \ldots \tilde{r}^n, 0$ ranged along the main diagonal (where r^j is as found in Eq. (49)), and $\Sigma_{0U} \triangleq [0U0U \ldots U]^T$ and $\Sigma_{U0} \triangleq [U0U0 \ldots 0]^T$ expressed in terms of the 3×3 null matrix 0 and the 3×3 unit matrix U.

Taking Laplace transforms of Eqs. (95) and (96) and solving for $\theta(s)$ in terms of $T(s)$ provides

$$\theta(s) = \left[I s^2 - s^4 \sum_{j=1}^{N} \frac{\delta_j^T \delta^j}{(s^2 + 2\zeta_j \sigma_j s + \sigma_j^2)} \right]^{-1} T(s) \,. \qquad (98)$$

When the principal axes of inertia are also the axes of attitude control, so I is diagonal, then only the appendage vibration terms found in Eq. (98) can couple the three scalar equations implied by this matrix equation. If this coupling is not too great, one can obtain useful information from scalar transfer functions typified by

$$\theta_\alpha(s) = G_\alpha(s)T_\alpha(s) \qquad \alpha = 1,2,3 \tag{99}$$

where

$$G_\alpha(s) = \sum_{j=1}^{N} \frac{s^2 + 2\zeta_j\sigma_j s + \sigma_j^2}{I_\alpha s^2[s^2(1 - \delta_\alpha^j\delta_\alpha^j/I_\alpha) + 2\zeta_j\sigma_j s + \sigma_j^2]} \, . \tag{100}$$

When $T_\alpha(s)$ consists of a reference torque $T_\alpha^R(s)$ plus a control torque $H_\alpha(s)\theta_\alpha(s)$, the system transfer function becomes

$$\theta_\alpha(s) = \left[\frac{G_\alpha(s)}{1 + G_\alpha(s)H_\alpha(s)}\right]T_\alpha^R(s) \, . \tag{101}$$

The characteristic roots of the system are then obtained as those values of s satisfying

$$1 + G_\alpha(s)H_\alpha(s) = 0 \, . \tag{102}$$

The practical application of these procedures is illustrated in Reference 17, a portion of which is reproduced here. This paper provides two examples, one sufficiently simple to provide physical insight into results, and the second representative of an actual spacecraft design.

Example: An Idealized Test Vehicle

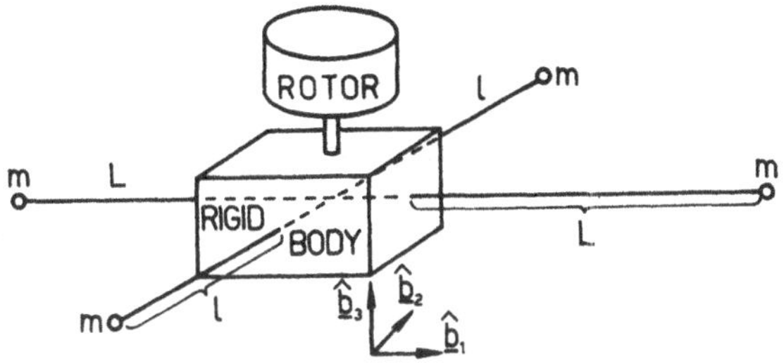

Fig. 3

Figure 3 shows four elastic memebers attached to a rigid body B which may also contain a rigid, symmetric rotor. The elastic bodies are massless except for a tip mass on each, normalized to unity (weight 1 lb). The central body B (including the rotor) weights 4 lbs., so the vehicle mass is distributed equally between the rigid part and the four flexible attachments, which together comprise what will be called a single appendage. The tip masses are particles, and the central body is inertially spherical, with weight moment of inertia 2590 lb-in^2 and dimensions small relative to beam lengths L and ℓ chosen as 4 ft and 2 ft, respectively. Beam stiffness are chosen to make the long beams have natural frequencies of 0.90 and 1.81 Hz , respectively, in the 1-2 and 1-3 planes of the vehicle when attached to a stationary base . The corresponding "cantilever mode" natural frequencies of the short beams are 2.71 and 4.52 Hz in the 1-2 and 2-3 planes, respectively. For the investigations reported in this paper, all beams are assumed longitudinally inextensible.

The modal coordinates in the matrix $\bar{\eta}$ of Eqs. (95) and (96) are not the cantilever modal coordinates of the individual beams, nor are they normal-mode coordinates of the total vehicle. (The presence of the rotor makes the vehicle normal-modes complex, and these modes are not used here in order to confine the analysis to real numbers.) The coordinates in $\bar{\eta}$ establish the response of the appendage in modes in which they would vibrate freely if the rigid body was free in translation but constrained against rotation. With this interpretation, it becomes evident that there must be eight coordinates in the column matrix η , and for this simple system one can almost guess the mode shapes.

Using established digital computer eigenvalue-eigenvector routines, one can determine the eight natural frequencies and mode shapes quite accurately (*). Results are presented schematically in Fig. 4, and quantitatively in Table 1. In the latter, node numbers and numbers associated with "d.o.f" (degree of freedom) are to be obtained from the first diagram of Fig. 4.

For purposes of preliminary design, it may be desirable to make exploratory transfer function calculations before

(*) Appreciation is expressed to Mr. John Garba of JPL, who actually performed these computer operations, using the SAMIS program.

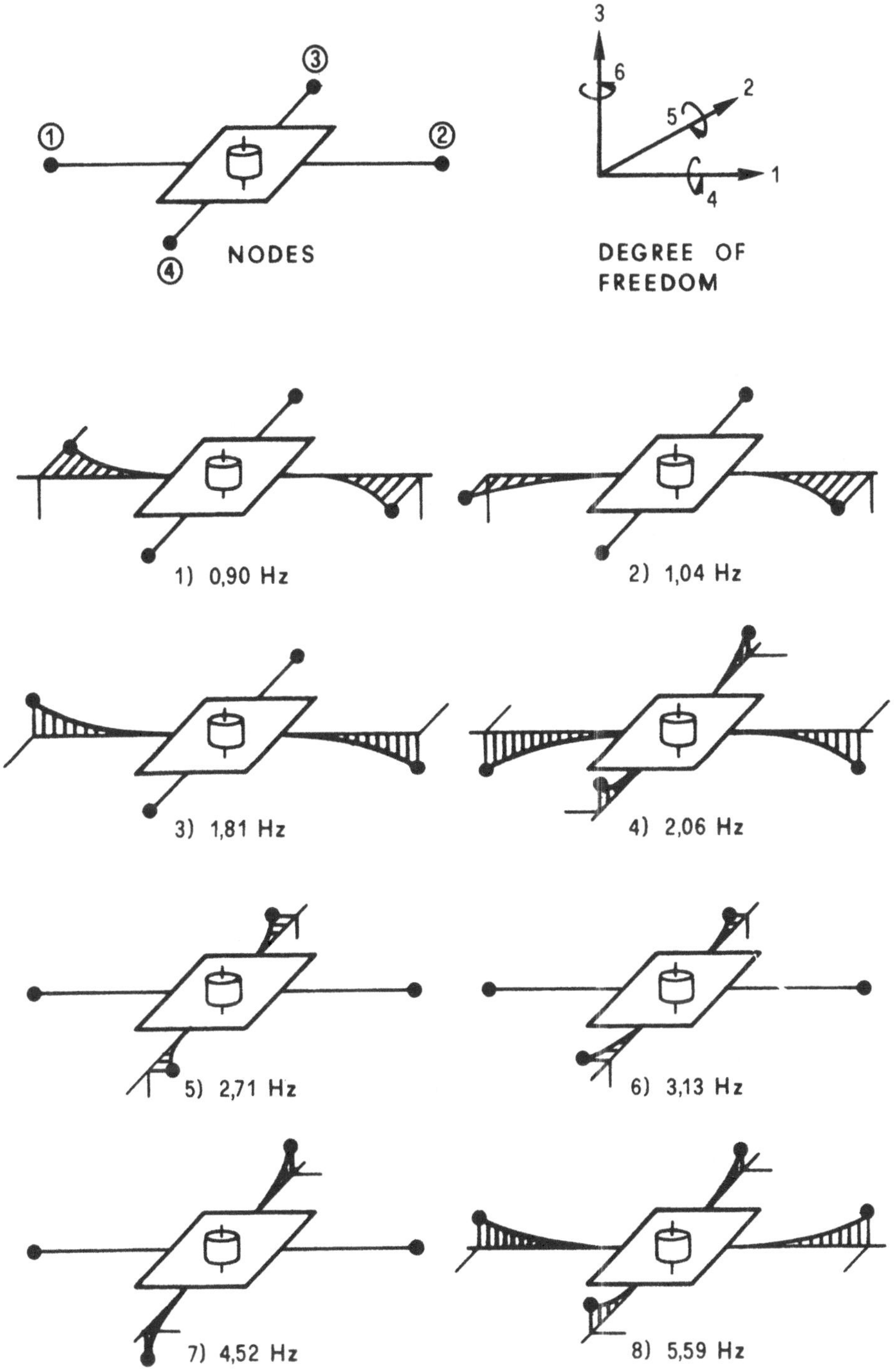

①
②
③
④
NODES
3
6
2
5
1
4
DEGREE OF FREEDOM
1) 0,90 Hz
2) 1,04 Hz
3) 1,81 Hz
4) 2,06 Hz
5) 2,71 Hz
6) 3,13 Hz
7) 4,52 Hz
8) 5,59 Hz

Fig. 4

Table 1. Mode shapes (M1–M8, see Fig. 4) and frequencies

Node/d.o.f.	Mode				
	M1	M2	M3	M4	M8
1/2	0.707	–0.817	0	0	0
1/3	0	0	0.707	–0.799	0.335
1/5	0	0	0.022	–0.025	0.010
1/6	–0.022	0.026	0	0	0
2/2	–0.707	–0.817	0	0	0
2/3	0	0	–0.707	–0.799	0.335
2/5	0	0	0.022	0.025	–0.010
2/6	–0.022	–0.026	0	0	0

Node/d.o.f.	M4	M5	M6	M7	M8
3/1	0	–0.707	–0.817	0	0
3/3	0.050	0	0	0.707	0.865
3/4	0.003	0	0	0.044	0.054
3/6	0	0.044	0.051	0	0
4/1	0	0.707	–0.817	0	0
4/3	0.050	0	0	–0.707	0.865
4/4	–0.003	0	0	0.044	–0.054
4/6	0	0.044	–0.051	0	0

the appendage is fully defined and subjected to modal analysis. Examination of the matrix transfer function in Eq. (9) reveals that of the appendage properties only the matrices $\bar{\sigma}$, $\bar{\zeta}$, and $\bar{\delta}$ are required.

For this simple system one can readily estimate $\sigma_1 \cong 0.9$ Hz, since this is the input cantilever mode natural frequency of the longest beams. Furthermore, one can utilize the physical interpretation of $\bar{\delta}^T\bar{\delta}$ and guess that in the first mode only the long beams will participate (as in Fig. 4); $\delta_3^j \delta_3^j$ should therefore be given by $2(1\ lb)(48\ in)^2 \cong 4600\ lb\text{-}in^2$, since this figure represents the contribution to I_3 of the two long beams. One might for this simple system easily guess as well the third, fifth, and seventh mode frequencies and contributions to $\bar{\delta}^T\bar{\delta}$, since these are asymmetric modes which have frequencies corresponding to the known cantilever mode frequencies. The physical interpretation of $\bar{\delta}^T\bar{\delta}$ tells us that the symmetric modes (modes 2, 4, 6, and 8 in Fig. 4) contribute nothing to this matrix, so they can be completely ignored.

Confirmation of these estimates for frequencies and $\delta^{j^T}\delta^j$ contributions comes from the computer-generated value of $\bar{\delta}$, using the definition following Eq. (96). The results for δ and $\delta^T\delta$ (with no truncation) are given by

$$\boldsymbol{\delta} = \begin{bmatrix} 0 & 0 & -2a \\ 0 & 0 & 0 \\ 0 & 2a & 0 \\ 0 & 0 & 0 \\ 0 & 0 & 0 \\ 0 & 0 & 0 \\ a & 0 & 0 \\ 0 & 0 & 0 \end{bmatrix} \qquad \boldsymbol{\delta}^T\boldsymbol{\delta} = \begin{bmatrix} b & 0 & 0 \\ 0 & 4b & 0 \\ 0 & 0 & 5b \end{bmatrix}$$

where $a = 33.94\ \ell b - in^2$; $b = 1150\ \ell b - in^2$; and computer generated numbers below 10^{-5} have been replaced by zeros. (For unit consistency it must be noted that $\boldsymbol{\delta}^T\boldsymbol{\delta}$ is actually $\bar{\boldsymbol{\delta}}^T E^{-1}$, with E a unit inertia matrix.) For this simple system, the estimated values are entirely correct. (The predicted $\boldsymbol{\delta}_3^1\boldsymbol{\delta} = 4600\ \ell b\text{-}in.^2$ checks -67.88 squared, and the calculated $\boldsymbol{\delta}^T\boldsymbol{\delta}$ is precisely the contribution of the appendage masses to the vehicle inertia matrix. Furthermore, the even-numbered rows of $\boldsymbol{\delta}$ are null, so these modes do not contribute to $\boldsymbol{\delta}^T\boldsymbol{\delta}$.)

Having obtained, by estimation or calculation, the necessary parameters of the flexible vehicle, the control system designer can begin the sequence of response analyses required for selection of a control system. In general, the first step might be the rapid construction of a number of root-locus plots for single-axis response, the second step might then be a series of eigenvalue studies for a smaller range of preliminary control system designs, and the final step would be a numerical integration of the most general system of equations of motion, to confirm the final control system design. For the simple test vehicle treated in this section, however, each appendage modal vibration contributes to the vehicle response about one axis only, so the three scalar equations of vehicle motion are coupled only by the "gyroscopic coupling" of the rotor. Furthermore, response θ_3 about the $\hat{\underline{b}}_3$ axis (see Fig. 3) is completely uncoupled from responses θ_1 and θ_2 about axes $\hat{\underline{b}}_1$ and $\hat{\underline{b}}_2$, so root-locus plots for this axis must indicate precisely the same response that would be obtained from eigenvalue calculations or numerical integrations. The purposes of this paper are best served by restricting the test vehicle analysis results to the presentation of root-locus plots for θ_3. In the next section eigenvalue analyses and integration results are presented for a space vehicle of realistic complexity.

Figures 5-8 portray root-locus plots for θ_3 response of the test vehicle, under various assumptions. The dashed-line loci on Figs. 5-7 are based on the assumption that the vehicle is rigid, so they provide for three simple control systems a basis for evaluating the influence of flexibility. Figures 5-7 introduce flexibility in the first mode of vibration (see Fig. 4 and Table 1). In this mode the normalized reduced inertia $\mathcal{R}_3^1$ is $1 - (\delta_3^1 \delta_3^1 / I_3) = 1 - [4600/(2590 + 5750)] = 0.45$. The natural frequency σ_1 is 0.90 Hz, although in the plots this value is normalized to unity. The damping ratio ζ_1 is assumed to be 0.05

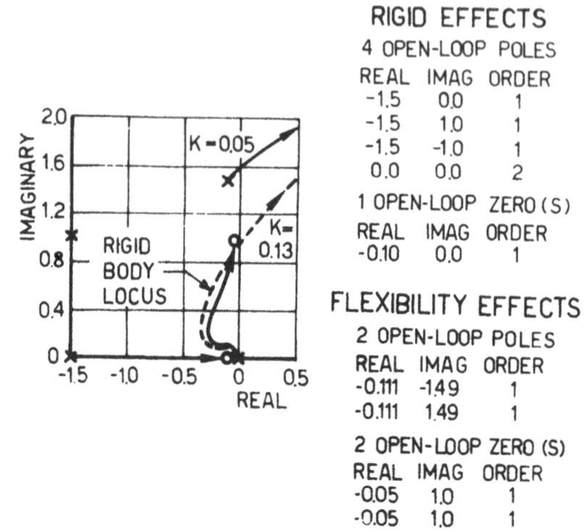
RIGID EFFECTS
4 OPEN-LOOP POLES
REAL IMAG ORDER
-1.5 0.0 1
-1.5 1.0 1
-1.5 -1.0 1
0.0 0.0 2
1 OPEN-LOOP ZERO (S)
REAL IMAG ORDER
-0.10 0.0 1
FLEXIBILITY EFFECTS
2 OPEN-LOOP POLES
REAL IMAG ORDER
-0.111 -1.49 1
-0.111 1.49 1
2 OPEN-LOOP ZERO (S)
REAL IMAG ORDER
-0.05 1.0 1
-0.05 1.0 1
IMAGINARY
REAL
K = 0.05
K = 0.13
RIGID BODY LOCUS
2.0
1.6
1.2
0.8
0.4
0
-1.5
-1.0
-0.5
0
0.5

Fig. 5

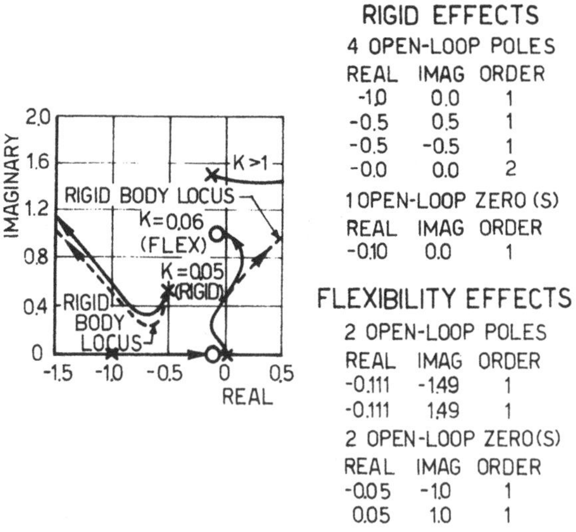
RIGID EFFECTS
4 OPEN-LOOP POLES
REAL IMAG ORDER
-1.0 0.0 1
-0.5 0.5 1
-0.5 -0.5 1
-0.0 0.0 2
1 OPEN-LOOP ZERO (S)
REAL IMAG ORDER
-0.10 0.0 1
FLEXIBILITY EFFECTS
2 OPEN-LOOP POLES
REAL IMAG ORDER
-0.111 -1.49 1
-0.111 1.49 1
2 OPEN-LOOP ZERO(S)
REAL IMAG ORDER
-0.05 -1.0 1
0.05 1.0 1
IMAGINARY
REAL
K >1
RIGID BODY LOCUS
K=0.06 (FLEX)
K=0.05 (RIGID)
RIGID BODY LOCUS
2.0
1.6
1.2
0.8
0.4
0
-1.5
-1.0
-0.5
0
0.5

Fig. 6

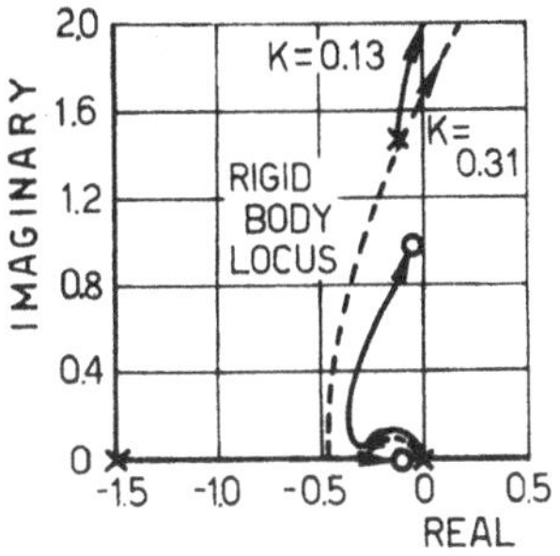

RIGID EFFECTS

3 OPEN-LOOP POLES

REAL	IMAG	ORDER
-1.5	0.0	1
-2.0	0.0	1
0.0	0.0	2

1 OPEN-LOOP ZERO (S)

REAL	IMAG	ORDER
-0.10	0.0	1

FLEXIBILITY EFFECTS

2 OPEN-LOOP POLES

REAL	IMAG	ORDER
-0.111	-1.49	1
-0.111	1.49	1

2 OPEN LOOP ZERO(S)

REAL	IMAG	ORDER
-0.05	-1.0	1
0.05	1.0	1

Fig. 7

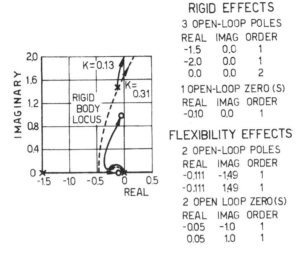

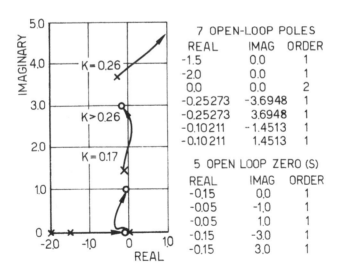

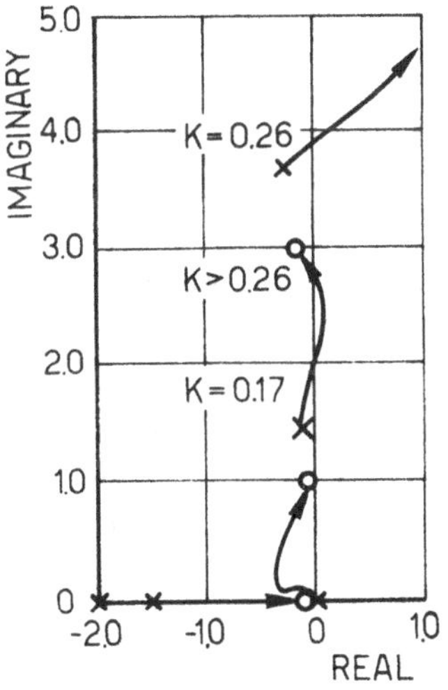

7 OPEN-LOOP POLES

REAL	IMAG	ORDER
-1.5	0.0	1
-2.0	0.0	1
0.0	0.0	2
-0.25273	-3.6948	1
-0.25273	3.6948	1
-0.10211	-1.4513	1
-0.10211	1.4513	1

5 OPEN LOOP ZERO (S)

REAL	IMAG	ORDER
-0.15	0.0	1
-0.05	-1.0	1
-0.05	1.0	1
-0.15	-3.0	1
-0.15	3.0	1

Fig. 8

(a relatively high figure is chosen to improve the visual impact of the root-locus plot). From Eqs. (100), the poles and zeros of the open-loop transfer function $G_3(s)$ may be obtained. With σ_1 normalized to unity, these roots become

$$p_{1,2} = 0,0 \qquad z_{1,2} = -0.05 \pm i$$

$$p_{3,4} = -0.111 \pm i1.49 \ .$$

For Fig. 5 the controller (feedback) transfer function $H_3(s)$ has a pair of complex poles, a real zero and a real pole of larger magnitude (so it might be called a lead-lag system with complex poles). The influence of flexibility on the locus is not surprising, but the "crossover gain" which marks the transition to instability is drastically reduced by the flexible appendage. Whereas a rigid vehicle would become unstable when the gain reaches 0.13, Fig. 5 shows instability to result when the gain becomes 0.05. This is a reduction for the flexible vehicle to 38% of the crossover gain for the rigid vehicle.

Parenthetically, it may be noted that the gain K employed in these root-locus plots is related to the gain K' which appears as a factor in the feedback transfer function by a factor which from Eq. (100) is uninfluenced by the flexibility of a single mode. Thus changes in root-locus plot gain K due to flexibility can be interpreted directly as changes in the actual feedback controller gain K'.

Figure 6 is, like Fig. 5, based on a lead-lag con

trol system with a pair of complex poles. The poles of $H_\alpha(s)$ in Fig. 6 are somewhat closer to the origin than in the previous example. The crossover gain for the flexible vehicle appears to be 20% higher than for the rigid vehicle, whereas this gain is 62% lower when the complex pole of $H_\alpha(s)$ matches the structural frequency, as in Fig. 5. Thus generalizations regarding the influence of appendage flexibility on control system stability would appear to be dangerous and superficial physical interpretations unwise.

Figure 7 portrays the rigid vehicle response to a "lead-lag" controller, with a negative real zero and two more remote negative poles (one of which is off the plot at -2.0). The crossover gain is $K = 0.31$. Figure 7 illustrates the influence of the first mode flexibility of the appendage, which reduces the crossover gain to 0.13, 42% of its original value. Figure 8 indicates the influence of the first and second modes of appendage vibration in combination.

For a "pure gain" direct feedback control, root-locus plots (not shown) can be deduced by inspection. It can be established by use of the Routhian array that such control systems cannot be destabilized by appendage flexibility.

For this idealized test vehicle there were generated 35 root-locus plots, of which those preceding are typical. With a standardized program (written by E. H. Kopf and R. Mankovitz), each plot required only 10 to 40 sec of IBM 7094 time, de pending on the number of plots requested in a given run. For the single-axis linear response of a vehicle as simple as this test vehicle, the conclusions are quantitatively precise. For a vehicle of realistic complexity (as treated in the next section), root-locus plots may lose quantitative validity, while preserving the qualitative significance normally sought in preliminary design studies.

8. Example: The TOPS autopilot

The unmanned TOPS is intended to perform scientific investigations of Jupiter, Saturn, Uranus, and Neptune in the late 1970's. Its configuration (Fig. 9) is dominated by a 14-ft-diam parabolic communications antenna and a bank of radioisotope thermoelectric generators (RTGs). In addition, a pair of telescoping booms are required to provide separation for sensitive instruments (magnetometers, plasma wave detector, etc.). Much of the remaining scientific instrumentation has been provided viewing area (around the antenna) and separation from RTG radiation by mounting it on a large foldout structure opposite the RTG foldout boom. Finally, central to the craft is an electronic equipment compartment carrying the autopilot's attitude sensors (high-gain gyros) and actuator (gimbaled engine). It is this portion of the vehicle that will be considered the rigid body to which the flexible appendages (RTG, science, and magnetometer booms; antenna) are attached.

While the hybrid coordinate formulation is also the ideal approach for the TOPS cruise attitude control system because of the use of momentum wheels, the dicussion here is limited to initial investigations of the autopilot which maintains vehicle attitude during the trajectory correction motor's thrusting periods. The gimbaled engine provides attitude control about the pitch and yaw axes, while roll control is maintained by the roll gas jets, which, in the cruise control mode, are normally used to desaturate the roll wheels. Pitch, yaw, and roll momentum wheels are switched out during the trajectory correction phase along with their respective cruise optical sensors. Rate and position sensing is provided in all three axes by high-gain rate-integrating gyros.

Preliminary efforts have attempted to determine the feasibility of using an autopilot control loop originally designed for a Mars orbital vehicle on TOPS as well. A linearized and simplified version of the proposed autopilot yaw axis (or pitch) loop is shown in Fig. 10. Among the assumptions made to develop the simple, single-axis system model was that gimbaled engine reaction torques on the craft would not be considered, i. e., the engine is assumed massless. Also, as discussed previously, the influence on θ (pitch or yaw) of constant thrust F_0 along the vehicle roll axis is ignored; only the effects of the torque applied by the gimbaled motor and the resulting rigid-flexible

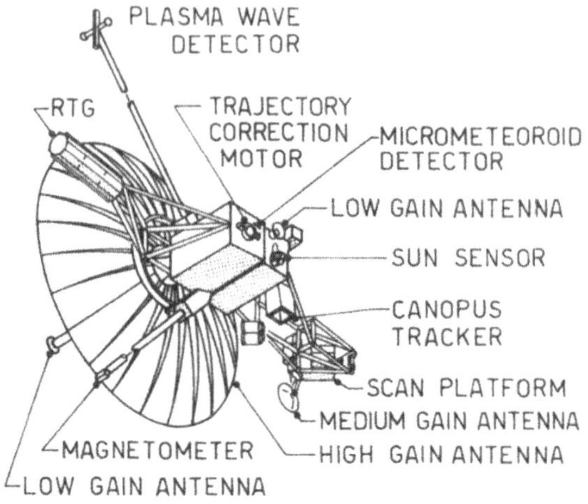
PLASMA WAVE DETECTOR
RTG
TRAJECTORY CORRECTION MOTOR
MICROMETEOROID DETECTOR
LOW GAIN ANTENNA
SUN SENSOR
CANOPUS TRACKER
SCAN PLATFORM
MEDIUM GAIN ANTENNA
HIGH GAIN ANTENNA
MAGNETOMETER
LOW GAIN ANTENNA

Fig. 9

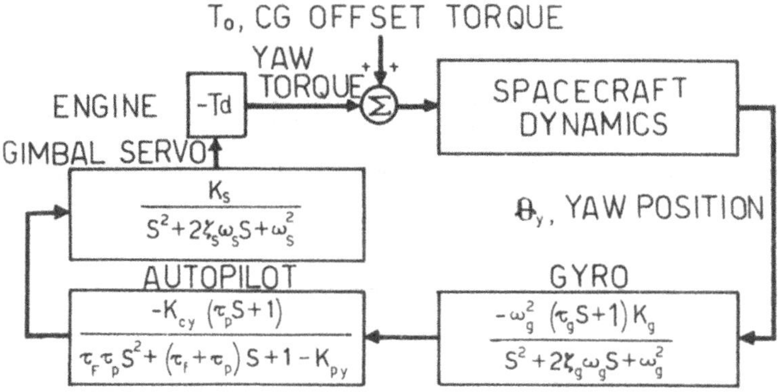
T₀, CG OFFSET TORQUE
YAW TORQUE
ENGINE GIMBAL SERVO
-Td
SPACECRAFT DYNAMICS
θy, YAW POSITION
AUTOPILOT
GYRO

Fig. 10

body interactions are included.

Of course, the three-stage design process to be
executed for the TOPS autopilot system must be based on the ex-
istence of a detailed hybrid coordinate structural model of the
craft pictured in Fig. 9 Such a model was developed (*) relatively quickly (2 weeks) using 20 discrete sub-bodies to approxi-
mate the flexible members. The result of computer calculations to
transform the discretized structural model into hybrid coordinate
modal data is summarized, in part in Table 2. Shown are the first
ten modal frequencies obtained and corresponding elements of the
matrices δ and $\phi^T M \Sigma_{E0}$. Approximately 30 modes were available in
all from the computer generated data.

It is at this stage that one must begin to weigh
the importance of the modes and, tentatively at least, truncate
the modal data. The first five modes, grouped near 1 Hz may be
traced to the antenna and magnetometer booms. One can see from
the size of coefficients in δ that pitch and yaw axes are predom
inantly affected by modes 1 and 4 while roll largely sees modes
2, 3, and 5. Based on very simple modeling of the craft as a rig
id body hinged to two rigid booms, natural frequencies of vibration above 4 Hz were shown to have little effect on autopilot

(*) Thanks must go to Messrs. Robert Bamford and Craig Helberg,
who modeled the structure and performed the necessary computer calculations.

Table 2. TOPS hybrid-coordinate structural model data

Mode i	σ_i, Hz	$\Phi^T M \Sigma_{E0}$ slug-ft			δ slug-ft^2		
		x	y	z	x	y	z
1	0.74	0.6381E-4	0.7994E-4	0.4092E00	0.6989E01	0.1150E02	0.3856E-2
2	0.75	0.3130E00	0.3519E00	-0.1537E-3	-0.3933E00	0.3582E00	0.1329E02
3	0.76	0.2964E00	-0.3620E00	-0.1059E-4	0.4072E00	0.3357E00	0.1254E02
4	0.76	-0.3970E-4	-0.4834E-4	0.5278E00	-0.8966E01	0.8642E01	-0.1440E-3
5	1.16	0.2777E-1	0.1248E-2	0.1468E-2	-0.2132E-1	0.1469E-1	0.4602E01
6	3.85	-0.2733E01	-0.30035E-1	-0.1554E-2	0.1673E00	0.6115E00	-0.2636E02
7	5.02	0.4006E00	0.4416E-1	0.2425E-1	-0.6592E00	0.4022E01	0.1107E01
8	5.66	-0.4516E-2	-0.2416E-2	-0.1875E00	0.8579E00	0.1944E01	-0.7297E-1
9	5.66	0.2074E00	0.1776E00	-0.7461E-2	-0.1245E00	0.2310E00	0.2673E01
10	5.69	-0.1780E-1	0.9784E-2	0.2416E00	-0.1100E01	0.1477E01	-0.2134E00

$$I = \begin{bmatrix} 1139.0 & -1.69 & -2.04 \\ -1.69 & 347.9 & -25.8 \\ -2.04 & -25.8 & 1264.0 \end{bmatrix} \text{slug-ft}^2, \quad \mathcal{M} = 38.54 \text{ slugs}$$

stability. Therefore modes beyond 5 were removed from the dynamic model for all phases of the analysis including the detailed computer simulation. An additional excuse for deleting the significantly higher frequency modes is the dramatic improvement in numerical integration speed if digital simulation is employed.

Equation 101 can now be of use in examining the linearized, single-axis control loop. For both pitch and yaw, modes 1 and 4 appear to dominate in the δ matrix, so that Θ_y and Θ_x could be approximated by equations of the form

$$\frac{\Theta_y(s)}{T_y(s)} = \left[I_{yy} s^2 - s^4 \frac{\delta_y^1 \delta_y^1}{(s^2 + 2\zeta_1 \sigma_1 s + \sigma_1^2)} - s^4 \frac{\delta_y^4 \delta_y^4}{(s^2 + 2\zeta_4 \sigma_4 s + \sigma_4^2)} \right]^{-1} .$$

(The equation for $\Theta_x(s)/T_x(s)$ is identical in form).
With the substitutions $I_{xx}, I_{yy}, \delta_y^1, \delta_x^1, \delta_y^4, \delta_x^4$ = 1139.0, 347.9, -11.5, -6.989, 8.642, and -8.966 slug - ft^2, respectively, and σ_1, σ_4 = 4.65 and 4.76 rad/sec, respectively, and $\zeta_1, \zeta_4 = 0.005$, the resulting transfer functions for yaw and pitch are

$$\frac{\Theta_y(s)}{T_y(s)} = \frac{(s^2 + 0.0465s + 21.62)(s^2 + 0.0476s + 22.8)}{347.9s^2[0.405s^4 + 0.066s^3 + 31.1s^2 + 2.09s + 493]}$$

$$\frac{\theta_x(s)}{T_x(s)} = \frac{(s^2 + 0.0465s + 21.62)(s^2 + 0.0476s + 22.8)}{1139s^2[0.887s^4 + 0.089s^3 + 41.9s^2 + 2.09s + 493]}$$

The open-loop poles and zeros of the linearized, single-axis autopilot system of Fig. 9 may now be plotted as shown in Figs. 11 and 12. The dashed lines on Figs. 11 and 12 indicate the closed-loop root locations under the ideal, totally rigid spacecraft condition where the control parameters are nominally specified as: $\tau_G = 1.77$ sec, $\tau_F = 0.111$ sec, $\zeta_G = 0.35$, $K_P = 2.2$, $\omega_G = 88.0$ rad/sec, $\zeta_s = 3.47$, $\tau_T = 20.0$ sec, and $\omega_s = 138.2$ rad/sec; gimbal servo poles: -938.66, - 20.35; gyro poles: -30.8 ± 82.4 ; autopilot poles: -9.12, +0.059; gyro zero: -0.565; and path guidance zero: -0.05.

Loop gain (DC or Bode gain) at the point of marginal stability is 10.0. On the other hand, in Fig. 11 the yaw axis loop with approximated flexible spacecraft dynamics shows a drastically changed root-locus with a critical gain reduced to 2.9. While the pitch-axis root-locus for the flexible case also departs significantly from the rigid case, critical gain reduction is only to 8.5. The fact that the total pitch axis moment of inertia is about 3.3 times as great as yaw's obviously was responsible for the difference in location of the flexible appendage pole-zero contributions to each axis and the effect on critical loop gain values.

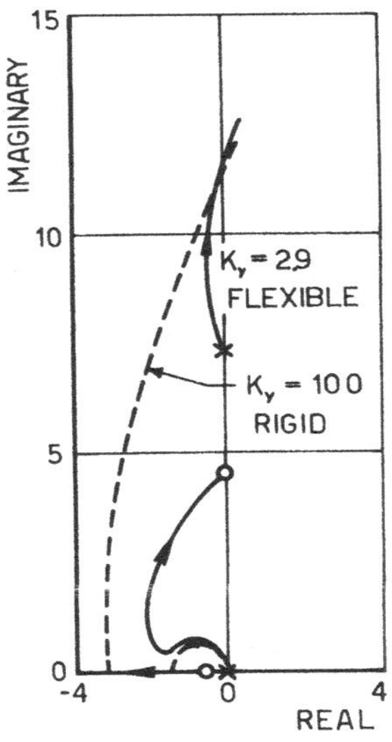

RIGID EFFECTS

7 OPEN-LOOP POLES

REAL	IMAG	ORDER
0.059	0.0	1
-9.12	0.0	1
-20.35	0.0	1
-938.7	0.0	1
-30.8	82.43	1
-30.8	-82.43	1
0.0	0.0	2

2 OPEN-LOOP ZERO (S)

REAL	IMAG	ORDER
-0.05	0.0	1
-0.565	0.0	1

FLEXIBILITY EFFECTS

4 OPEN-LOOP POLES

REAL	IMAG	ORDER
-0.0579	-7.37	1
-0.0579	7.37	1
-0.0237	-4.73	1
-0.0237	4.73	1

4 OPEN LOOP ZERO (S)

REAL	IMAG	ORDER
-0.0239	-4.77	1
-0.0239	4.77	1
-0 0232	-4.64	1
-0 0232	4.64	1

Fig. 11

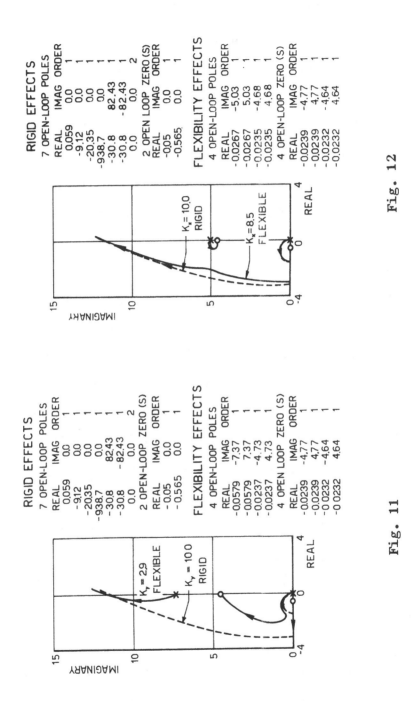

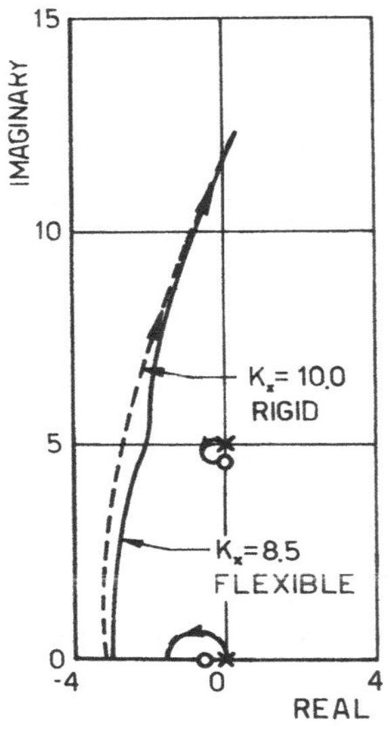

RIGID EFFECTS

7 OPEN-LOOP POLES

REAL	IMAG	ORDER
0.059	0.0	1
-9.12	0.0	1
-20.35	0.0	1
-938.7	0.0	1
-30.8	82.43	1
-30.8	-82.43	1
0.0	0.0	2

2 OPEN LOOP ZERO (S)

REAL	IMAG	ORDER
-0.05	0.0	1
-0.565	0.0	1

FLEXIBILITY EFFECTS

4 OPEN-LOOP POLES

REAL	IMAG	ORDER
-0.0267	-5.03	1
-0.0267	5.03	1
-0.0235	-4.68	1
-0.0235	4.68	1

4 OPEN-LOOP ZERO (S)

REAL	IMAG	ORDER
-0.0239	-4.77	1
-0.0239	4.77	1
-0.0232	-4.64	1
-0.0232	4.64	1

Fig. 12

An examination of the rigid-body autopilot system reveals closed-loop bandwidths of about 0.3 and 0.8 corresponding to DC loop gains of 1.0 (0 dB) and 3.16 (10 dB), respectively. It is clear that the first four appendage modes have already entered the control bandwidth at the latter gain level and, in the case of yaw, caused instability.

One might expect that, if additional modes are used in the initial single-axis root-locus analyses, critical gain values would decrease even more. Further, the fact that interaxis coupling might require even more substantial gain reductions to maintain stability suggests an eigenvalue analysis of the complete system. The TOPS autopilot may be arranged into state variable form to give a 30×30 system state matrix including pitch, yaw, and roll angles, five flexible appendage modes, and the various control dynamics. Figure 13 shows that when the detailed, coupled case is considered, yaw loop stability deteriorates further and essentially controls the stability of the entire system.

A digital simulation was programmed and included nonlinearities due to 1) gimbal servo drive amplifier saturation, 2) the saturation characteristic built into the compensation block to prevent reaching mechanical gimbal stops, and 3) the roll axis bang-bang control loop. (It did not include gimbal actuator stiction.) The level of detail was such as to include effects of gimbal mounting errors and center-of-mass displacements due to

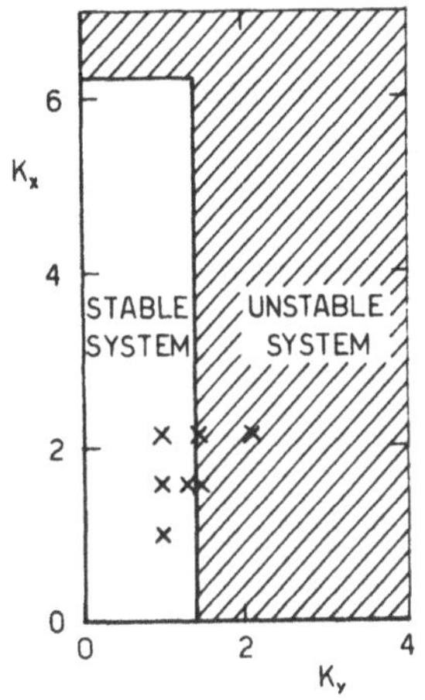

Fig. 13

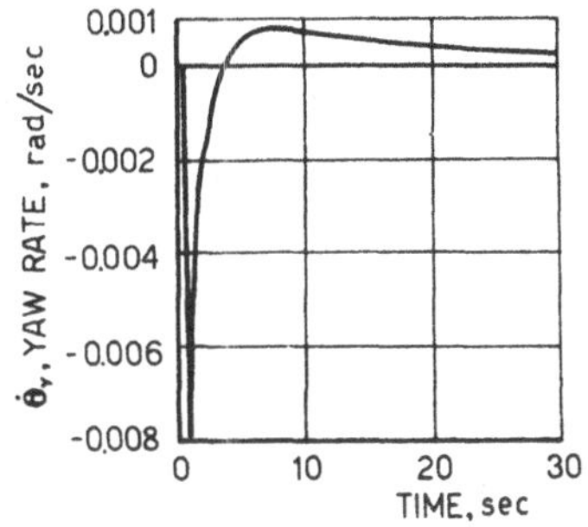

Fig. 14

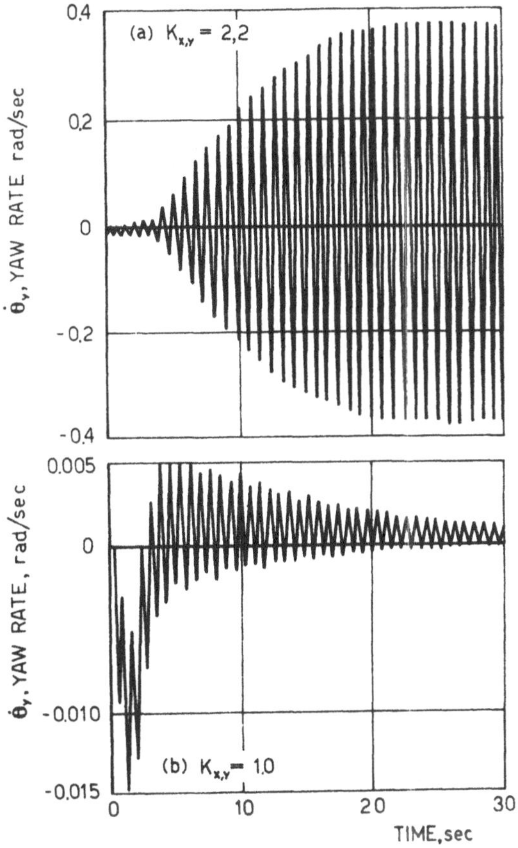

Fig. 15

appendage deformations and the hybrid coordinate formulation represented by Eqs. (1) and (2) was implemented in detail. Yaw-axis response is shown for three cases in Figs. 14 and 15 to illustrate the computer simulation results. In each case, the autopilot controller and spacecraft are responding to the engine turn-on transient (100-lb thrust) under the conditions of an initial pitch axis center-of-mass offset of 0.02 ft. Five flexible modes are used in each case in Fig. 15; for gains of 2.2 the flexible vehicle proves to be unstable (Fig. 15a) as the eigenvalue analysis predicted. Reduction of $K_{x,y}$ to 1.0 results in a stable response (Fig. 15b), which also agrees with eigenvalue predictions.

9. Liapunov stability analysis of spinning flexible spacecraft

In this section the results of a recent study reported in Ref. 19 are briefly reviewed. The objective of the paper is a set of stability criteria for a spacecraft characterized as a rigid body having attached a flexible appendage idealized as a collection of elastically interconnected particles. The appendage is permitted in the paper to be configured in a general fashion, but the special case of the planar appendage shown in Fig. 16 is adopted in order to obtain explicit closed form stability criteria without the aid of computer simulation. The well known stability procedures offered by Liapunov's direct (second) method are invoked in the paper. Among the many theorems on stability and instability of the class developed by Liapunov two have direct relevance to the development. Described in detail by Pringle, [20] the theorems of interest are paraphrased below.

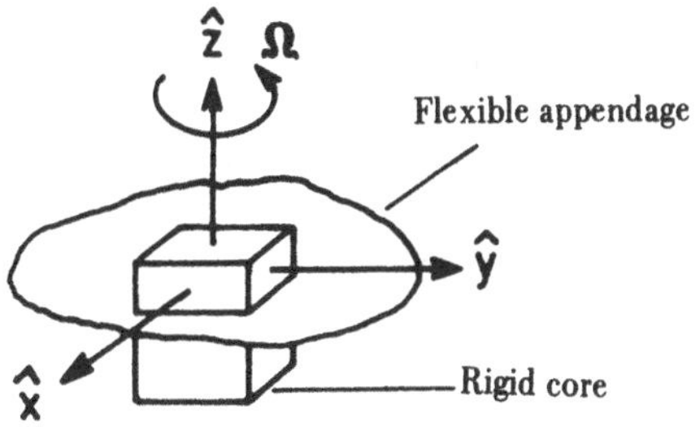

Fig. 16 Restricted Appendage Model

Theorem 1: The null solution $X(t) = 0$ of the differential equation $\dot{X} = F(X)$ is asymptotically stable if there exists a function $L(X)$ in a region around the origin both positive definite and strictly decreasing for all solutions in that region except for $X \equiv 0$.

Theorem 2: The null solution $X(t) = 0$ of the differential equation $\dot{X} = F(X)$ is unstable if there exists a function $L(X)$ in a region around the origin both negative definite (or sign variable) and strictly decreasing for all solutions in that region except for $X \equiv 0$.

Although the implementation of Liapunov's direct method is impeded by the lack of a general formal procedure for the generation of a testing function, the Hamiltionian serves this purpose for a wide class of dynamical systems. Specifically if the total energy of the system is free of explicit time dependence, then for completely damped systems the Hamiltonian, (*)

(*) The term Hamiltonian is here applied to the function ./.

H is a suitable testing function for asymptotic stability and instability. For our purpose the concept "complete damping" requires that energy be dissipated for any possible motion other than the nominal motion in the neighbourhood of the nominal motion in the coordinate space adopted. However, the damping of a freely spinning body with internal energy dissipation is not complete in terms of inertial attitude angles which are zero only prior to perturbation, since after perturbation the vehicle cannot return to its original state. Thus for such systems the Hamiltonian is not strictly decreasing in the neighbourhood of the null solution, and therefore asymptotic stability cannot be proclaimed as a consequence of the positive definiteness of H. In 1969 R. Pringle [20] provided a method to circumvent this problem. The procedure is to constrain the attitude angles through the angular momentum integral in such a fashion that they represent deviations of the body spin axis from an inertial direction, $\hat{\underline{n}}_3$, which is colinear with the instantaneous angular momentum vector h before or after perturbation from its nominal inertial orientation. The resulting attitude angles (defined as θ_1 and θ_2 in the sequel) will in the case of a stable vehicle ultimately reduce to zero after initial perturbation, thus assuring complete damping and asymptotic stability. This assurance requires also a ju-

---

$H \equiv \sum_{i=1}^{n} \frac{\partial L}{\partial \dot{q}_i}\dot{q}_i - L$, where L is the Lagrangian and $q_i (i = 1,\dots,n)$ is a generalized coordinate of the system. This usage is not universal.

dicious choice of deformation variables, which must vanish whenever the vehicle adopts a steady-state configuration and rotates about the body spin axis (at whatever rate is consistent with the actual angular momentum).

The procedure then will be as follows:

1) Derive the Hamiltonian in terms of the inertial angular velocity components $\omega_x, \omega_y, \omega_z$ of the core body and a set of deformation variables u_x^i, u_y^i, u_z^i ($i = 1, \ldots, N$ for N particles in the flexible appendage).

2) Introduce the angular momentum integral defining $\underline{h}$ as the angular momentum after perturbation directed along the unit vector $\hat{\underline{n}}_3$.

3) Express $\underline{h}$ in the body fixed vector basis $\{\hat{\underline{b}}\} \triangleq \{\hat{\underline{b}}_1, \hat{\underline{b}}_2, \hat{\underline{b}}_3\}^T$ as a function of the inertial angular velocity components and the deformation variables. Define the 3×1 matrix $h_b \triangleq \{\hat{\underline{b}}\} \underline{h}$.

4) Introduce the transformation H which transforms the body vector basis $\{\hat{\underline{b}}\}$ into the inertially fixed vector basis $\{\hat{\underline{n}}\} = \{\hat{\underline{n}}_1, \hat{\underline{n}}_2, \hat{\underline{n}}_3\}^T$, where $\hat{\underline{n}}_3$ is indicated in step (2) and $\hat{\underline{n}}_1$ and $\hat{\underline{n}}_2$ are arbitrarily chosen within the constraints of the dextral orthonormal set in $\{\hat{\underline{n}}\}$.

5) From the three scalar Equations

$$\underline{h} \cdot \hat{\underline{n}}_1 = 0$$

$$\underline{h} \cdot \hat{\underline{n}}_2 = 0$$

$$\underline{h} \cdot \hat{\underline{n}}_3 = h$$

(where h is the magnitude of the perturbed angular momentum vector), and the transformation $\Theta h_b = (0,0,h)^T$ solve for the angular velocity components as functions of the deformation variables and the attitude angles (assumed small) comprising Θ.

6) Eliminate the angular velocity components from the Hamiltonian, thus expressing the Hamiltonian as a function of the attitude angles and the deformation variables.

7) Extract stability criteria by implementing Theorems 1 and 2, using coordinate transformations for $u_x^i, u_y^i, u_z^i (i = 1,\dots,n)$ where truncation of distributed coordinates is warranted and necessary for literal formulation of stability criteria.

As shown in Ref. 19, Steps 1 through 6 produce the Hamiltonian

$$H = \frac{1}{2}h^2\theta_2^2\frac{(A-C)}{AC} + \frac{1}{2}h^2\theta_1^2\frac{(C-B)}{BC}$$

$$-\frac{h^2\theta_2}{AC}\delta_x^T\eta + \frac{h^2\theta_1}{BC}\delta_y^T\eta$$

$$+\frac{1}{2}\eta^T\left[\omega^2 + \frac{h^2}{AC^2}\delta_x\delta_x^T + \frac{h^2}{BC^2}\delta_y\delta_y^T\right]$$

$$+\frac{1}{2}\dot{\eta}^T\left[U - \frac{1}{2}\delta_y\delta_y^T - \frac{1}{B}\delta_x\delta_x^T\right]\dot{\eta}\ .$$

Here A, B and C are principal axis moments of inertia of the total vehicle, δ_x and δ_y are (for an appendage with N modes of deformation permitted) N by 1 column matrices that embody the properties of mode shape and appendage mass, ω^2 is a diagonal N by N matrix of modal frequencies, η is an N by 1 matrix of modal deformation coordinates of the appendage, U is the N by N unit matrix, and (as previously noted) h is the angular momentum magnitude and θ_1, θ_2 are inertial attitude variables. If now we permit only one mode of deformation (so that $N = 1$, we can write the stability condition (H positive definite) as:

$$\frac{1}{2}h^2\theta_2^2\frac{(C-A)}{AC} + \frac{1}{2}h^2\theta_1^2\frac{(C-B)}{BC}$$
$$-\frac{h^2\theta_2}{AC}\delta_{x1}\eta_1 + \frac{h^2\theta_1}{BC}\delta_{y1}\eta_1$$
$$+\frac{1}{2}\eta_1^2\left(\omega_1^2 + \frac{h^2}{AC^2}\delta_{x1}^2 + \frac{h^2}{BC^2}\delta_{y1}^2\right)$$
$$+\frac{1}{2}\dot{\eta}_1^2\left(AB - B\delta_{y1}^2 - A\delta_{x1}^2\right) > 0$$

where it is to be understood that >0 means positive for all values of θ_1, θ_2, η_1 and $\dot{\eta}_1$ in the neighbourhood of the origin $\theta_1 = \theta_2 = \eta_1 = \dot{\eta}_1 = 0$, except equal to zero at the origin itself. Note that the last term in this expression, and similarly its general counterpart preceding, is uncoupled from the remaining terms; moreover by properties of physically realizable inertia matrices (positive definite) it can be shown that this term is positive. The stability condition then reduces to:

$$(\theta_1\theta_2\eta_1)\begin{pmatrix} \frac{h^2}{2}\frac{(C-B)}{BC} & 0 & \frac{h^2}{2}\frac{\delta_{y1}}{BC} \\ 0 & \frac{h^2}{2}\frac{(C-A)}{AC} & -\frac{h^2}{2}\frac{\delta_{x1}}{AC} \\ \frac{h^2}{2}\frac{\delta_{y1}}{BC} & -\frac{h^2}{2}\frac{\delta_{x1}}{AC} & \frac{1}{2}\left(\omega_1^2+\frac{h^2}{AC^2}\delta_{x1}^2+\frac{h^2}{BC^2}\delta_{y1}^2\right) \end{pmatrix}\begin{pmatrix}\theta_1\\ \theta_2\\ \eta_1\end{pmatrix} > 0.$$

The sign character of the above quadratic function is determined by testing the sign character of its corresponding symmetric matrix; and by Sylvester's Theorem we are assured that for the cited matrix to be positive definite it is necessary and sufficient that all principal diagonal minors be simultaneously positive. If this test fails, H is either sign variable (implying instability) or positive semidefinite (but not positive definite). If we exclude this latter class of systems (thereby excluding limiting casesaxis symmetric vehicles with $C = A$ or $C = B$), then conditions both necessary and sufficient for asymptotic stability of the restricted planar appendage model are:

$$\frac{h^2}{2}\frac{(C-B)}{BC} > 0$$

$$\frac{h^2}{2}\frac{(C-B)}{BC}\cdot\frac{h^2}{2}\frac{(C-A)}{AC} > 0$$

$$\frac{h^2}{2}\frac{(C-B)}{BC}\left\{\frac{h^2}{2}\frac{(C-A)}{AC}\left[\frac{\omega_1^2}{2}+\frac{h^2\delta_{y1}^2}{2AC^2}+\frac{h^2\delta_{y1}}{2BC^2}\right]-\frac{h^4\delta_{x1}^2}{4A^2C^2}\right\}$$
$$-\frac{h^2\delta_{y1}^2}{2BC}\left[\frac{h^4\delta_{y1}}{4BC}\frac{(C-A)}{AC}\right] > 0 .$$

The combination of the first two conditions, as predicted by energy sink methods, requires that the spin axis be the axis of maximum of inertia, i.e.,

$$C > A \quad \text{and} \quad C > B .$$

This, of course, we have observed before. In addition however a new criterion emerges, requiring satisfaction of the third condition above. After expansion and combination of terms this additional criterion takes the form

$$\omega_1^2 > \frac{h^2}{C^2}\left[\frac{\delta_{x1}^2(C-B)+\delta_{y1}^2(C-A)}{(C-A)(C-B)}\right].$$

By replying h by its zero th order approximation $C\Omega$ where Ω is the nominal spin frequency, the above condition simplifies to the following:

$$\left(\frac{\omega_1}{\Omega}\right)^2 > \frac{\delta_{x1}^2(C-B)+\delta_{y1}^2(C-A)}{(C-A)(C-B)} .$$

Thus a stability criterion arises which explicitly bounds the first modal frequency of the N particle structure.

Algebraic difficulty precludes the generation of explicit stability criteria for more than one mode; however, for reference, the general stability criterion is written in its simpler matrix form below.

$$(\theta_1 \theta_2 \vdots \eta_1 \eta_2 \ldots) \left[\begin{array}{cc|cccc} \frac{h^2(C-B)}{2CB} & 0 & \frac{h^2\delta_{y1}}{BC} & \frac{h^2\delta_{y2}}{BC} & \cdots \\ 0 & \frac{h^2(C-A)}{2AC} & -\frac{h^2\delta_{x1}}{AC} & -\frac{h^2\delta_{x2}}{AC} & \cdots \\ \hline \frac{h^2\delta_{y1}}{BC} & -\frac{h^2\delta_{x1}}{AC} & \frac{h^2\delta_{x1}^2}{2AC^2}+\frac{h^2\delta_{y1}^2}{2BC^2}+\frac{\omega_1^2}{2} & 0 & \cdots \\ \frac{h^2\delta_{y2}}{BC} & -\frac{h^2\delta_{x2}}{AC} & 0 & \frac{h^2\delta_{x2}^2}{2AC^2}+\frac{h^2\delta_{y2}^2}{2BC^2}+\frac{\omega_2^2}{2} & \\ \vdots & \vdots & \vdots & & \ddots \end{array}\right] \begin{pmatrix} \theta_1 \\ \theta_2 \\ \hline \eta_1 \\ \eta_2 \\ \vdots \end{pmatrix} > 0 .$$

Bibliography for lectures 1 – 9

[1] W.W. Hooker and G. Margulies, "The Dynamical Attitude Equations for an n-Body Satellite, " J.Astronaut. Sci., Vol. 12, 1965, pp. 123-128.

[2] F. Buckens, "On the Influence of the Elasticity of Components in a Spinning Satellite of Its Motion," Proc. of the 16th International Astronautical Congress (Athens, 1965).

[3] L. Meirovitch and H.D. Nelson, "On the High-Spin Motion of a Satellite Containing Elastic Parts," J. Spacecraft and Rockets, Vol. 3, 1966, pp. 1597-1602.

[4] R.E. Roberson and J. Wittenburg, "A Dynamical Formalism for an Arbitrary Number of Interconnected Rigid Bodies, with Reference to the Problem of Satellite Attitude Control," Proc. of the 3rd International Congress of Automatic Control, (London 1966), Butterworth and Co., 1967, pp. 46D.1 - 46D.8.

[5] H.Ashley, "Observations on the Dynamic Behaviour of Large, Flexible Bodies in Orbit," AIAA J. Vol. 5, 1967, pp. 460-469.

[6] R.D. Milne, "Some Remarks on the Dynamics of Deformable Bodies," AIAA J., Vol. 6, 1968, pp. 556-558.

[7] P.W. Likins and P.H. Wirsching, "Use of Synthetic Modes in Hybrid Coordinate Dynamic Analysis," AIAA J., Vol. 6, 1968, pp. 1867-1872.

[8] J.L. Farrell and J.K. Newton, "Continuous and Discrete RAE Models," J. Spacecraft and Rockets, Vol. 6, 1969, pp. 414-423.

[9] P.W. Likins, "Dynamics and Control of Flexible Space Vehicles," TR 32-1329, Jet Propulsion Laboratory, 1970.

[10] W.B. Gevarter, "Basic Relations for Control of Flexible Vehicles," AIAA J., Vol. 8, 1970, pp. 666-672.

[11] P.W. Likins and A.H. Gale, "Analysis of Interactions Between Attitude Control Systems and Flexible Appendages," Proc. of the 19th International Astronautical Congress (New York 1968), Vol. 2, Pergamon Press, 1970, pp. 67-90.

[12] W.W Hooker, "A Set of Dynamic Attitude Equations for an Arbitrary n-Body Satellite Having r Rotational Degrees of Freedom," AIAA J., Vol. 8, 1970, pp. 1205-1207.

[13] A.H. Gale and P.W. Likins, "Influence of Flexible Appendages on Dual-Spin Spacecraft Dynamics and Control," J. Spacecraft and Rockets, Vol.7 1970 pp 1049-1056.

[14] P.Y. Willems, "Stability of Deformable Gyrostats on a Circular Orbit," J. of the Astronaut. Sci., Sept.-Oct. 1970, XVIII-2, 65-85.

[15] F.R. Vigneron, "Stability of a Freely-Spinning Satellite of Crossed Dipole Configuration," Canadian Aeronautics and Space Institute Trans. Vol. 3, March 1970, pp. 8-19.

[16] P.W. Likins and H.K. Bouvier, "Attitude Control of Nonrigid Spacecraft," Astronautics and Aeronautics, Vol. 9, 1971, pp. 64-71.

[17] P.W. Likins and G.E. Fleischer, "Results of Flexible Spacecraft Attitude Control Studies Utilizing Hybrid Coordinates," J. Spacecraft and Rockets, Vol. 8, March 1971, pp. 264-273.

[18] P.W. Likins, "Finite Element Appendage Equations for Hybrid Coordinate Dynamic Analysis," International Journal of Solids and Structures, Vol. 8, 1972, pp. 709-731.

[19] F.S. Barbera and P.W. Likins, "Liapunov Stability Analysis of Spinning Flexible Spacecraft," to appear in A.I.A.A. Journal, March 1973.

[20] R. Pringle, Jr., "Stability of Force-Free Motions of a Dual-Spin Spacecraft," A.I.A.A. Journal, Vol. 7, June 1969, pp. 1054-1063.

Nomenclature for lecture notes 1 – 9

Latin Symbols

$\underline{A}, \bar{A}$ = inertial acceleration of element field point p, vector and (3x1) matrix in basis $\{\underline{e}\}$, respectively

$\underline{A}^j$ = inertial acceleration of node j

A = (12n x 12n) coefficient matrix in Eq. (71)

α = reference frame established by Q and $\underline{a}_1, \underline{a}_2, \underline{a}_3$

$\{\underline{a}\}$ = (3 x 1) vector array with elements $\underline{a}_1, \underline{a}_2, \underline{a}_3$ the dextral orthogonal unit vectors fixed in

B = (12n x 12n) coefficient matrix in Eq. (71)

b = reference frame established by base body

$\{\underline{b}\}$ = (3x1) vector with elements $\underline{b}_1, \underline{b}_2, \underline{b}_3$ the dextral orthogonal unit vectors fixed in b

C = (3 x 3) variable direction cosine matrix; $\{\underline{a}\} = C\{\underline{b}\}$

$\bar{C}^s, \bar{C}$ = (3 x 3) constant direction cosine matrix; $\{\underline{e}^s\} = \bar{C}^s\{\underline{a}\}$ and generic $\bar{C}^s$

CM = vehicle mass center

CM^s = s^{th} element mass center when appendage in steady-state

$\underline{c}, c$ = vector from CM to point O and (3x1) matrix in basis $\{\underline{a}\}$; respectively

D = displacement coefficient matrix in $\bar{\varepsilon} = D\bar{\omega}$ (Eq. (12))

E = Young's modulus

E = number of finite elements

E_j = set of numbers of finite elements in contact with node j

$\underline{\delta}, \delta$ = contribution to $\underline{c}$ not attributable to appendage deformation, Eq. (43), vector and (3×1) matrix in basis $\{\underline{a}\}$ respectively

$\{\underline{e}^s\}, \{\underline{e}\}$ = (3×1) vector array with elements $\underline{e}_1^s, \underline{e}_2^s, \underline{e}_3^s$ the dextral orthogonal unit vectors fixed in the sth finite element in its steady-state; and generic $\{\underline{e}^s\}$

F = $(6N\times 6N)$ matrix relating $\bar{y}$ to Γ, Eq. (9)

$\underline{F}^j, F^j$ = resultant force on j^{th} nodal body vector and (3×1) matrix in basis $\{\underline{a}\}$ respectively

$\underline{F}^{sj}, \bar{F}^{sj}$ = force applied by finite element s to nodal body j, vector and (3×1) matrix in basis $\{\underline{e}\}$, respectively

$\bar{F}^{js}$ = $-\bar{F}^{sj}$

$\underline{f}^j, f^j$ = resultant for nodal body j of forces external to the system vector, and (3×1) matrix in basis $\{\underline{a}\}$, respectively

$\bar{G}(\xi,\eta,\zeta)$ = (3×1) matrix function of element body forces in basis $\{\underline{e}\}$

G' = $(6N\times 6N)$ gyroscopic coupling matrix in Eq. (64)

$\bar{g}$ = element gyroscopic coupling matrix $(6N\times 6N)$

$\underline{H}^j$ = angular momentum of nodal body j for its mass center

$\underline{\underline{I}}^j, I^j$ = inertia dyadic of nodal body j for its mass center, and (3×3) inertia matrix in basis $\{\underline{n}^j\}$

$I_{\alpha\gamma}$ = element of I^j, $\alpha,\gamma = 1,2,3$

I = inertially fixed point

$\{\underline{i}\}$ = (3×1) vector array of inertially fixed, dextral, orthogonal, unit vectors $\underline{i}_1, \underline{i}_2, \underline{i}_3$

i = $(-1)^{1/2}$

K' = $(6N\times 6N)$ appendage stiffness matrix (Eq. (64))

$\bar{k}, k$ = $(6N \times 6N)$ finite element structural stiffness matrix, for vector bases $\{\underline{e}\}$ and $\{\underline{a}\}$ respectively

$\bar{k}_0$ = $(6N \times 6N)$ stiffness matrix for unloaded element for basis $\{\underline{e}\}$

$\bar{k}_\Delta$ = $(6N \times 6N)$ preload (geometric) stiffness matrix for element, basis $\{\underline{e}\}$

L = $(12N \times 1)$ matrix in Eq. (71)

L' = $(6N \times 1)$ matrix in Eq. (64)

$\bar{L}^s, \bar{L}$ = $(6N \times 1)$ matrix of forces and torques on the nodes of element s, Eq. (19); and generic $\bar{L}^s$

M' = $(6N \times 6N)$ generalized inertia matrix, Eq. (64)

$M, M^c, \bar{M}$ = constituents of M', Eq. (65)

M, M_s = vehicle mass; mass of finite element s

$\bar{m}, m$ = generic $(6N \times 6N)$ consistent mass matrix for finite element, bases $\{\underline{e}\}$ and $\{\underline{a}\}$ respectively, Eqs. (35), (40)

m_j^ℓ, m^j = mass of jth nodal body, and mass matrix $m^j = m_j U$

N = number of modal coordinates after truncation

N_s, N = number of nodes for finite element s, and genric N_s

n = number of nodes in appendage

$\{\underline{n}^j\}, \{\underline{n}\}$ = (3×1) vector array with elements $\underline{n}_1^j, \underline{n}_2^j, \underline{n}_3^j$ the dextral orthogonal unit vectors fixed in nodal body j, and generic form

O = point fixed in b, and vehicle CM for steady-state deformation

P = $(3 \times 6N)$ matrix relating $\bar{\omega}$ to Γ, Eq. (5)

Q = $(12N \times 1)$ state matrix, Eq. (71)

$\mathcal{Q}$ = point common to a and b

q = $(6N \times 1)$ matrix of variational deformation variables, Eq. (63)

$\underline{R}, R$ = vector from 0 to Q, and (3×1) matrix in basis $\{\underline{a}\}$

$\underline{R}_c^s, R_c^s$ = vector from Q to CM^s, and (3×1) matrix in basis $\{\underline{a}\}$ respectively

$\underline{r}^j, r^j$ = vector from Q to steady-state node j, and (3×1) matrix in basis $\{a\}$, respectively

S = (6×6) coefficient matrix in stress-strain Equation (15)

$\underline{t}^j, t^j$ = external torque on nodal body j, vector and (3×1) matrix in basis $\{\underline{a}\}$, respectively

$\underline{T}^j, T^j$ = torque on nodal body j, vector and (3×1) matrix in basis $\{\underline{a}\}$, respectively

$\underline{T}^{sj}, \bar{T}^{sj}$ = torque on nodal body j applied by elements s, vector and (3×1) matrix in basis $\{\underline{e}^s\}$ respectively

U = (3×3) unit matrix

U^* = virtual strain energy

$\underline{u}^j, u^j$ = displacement of node j due to variations from steady state deformation (i.e., variational translational nodal deformation), vector and (3×1) matrix in basis $\{\underline{a}\}$, respectively

W = $(3\times6N)$ matrix relating $\bar{\omega}$ to $\bar{y}$, Eq. (11)

w^* = virtual work

$\underline{w}^s, \bar{w}^s, w^s$ = displacement of field point of finite element s due to variations from steady-state deformation, (i.e., variational element deformation), vector and (3×1) matrices in basis $\{\underline{e}^s\}$ and $\{\underline{a}\}$, respectively

$\underline{w}, \bar{w}, w$ = generic for $\underline{\omega}^s$, $\bar{\omega}^s$ and ω^s

$\underline{X}, X$ = vector from I to CM, and (3×1) matrix in basis $\{\underline{i}\}$

$Y, \bar{Y}$ = $(12n \times 1)$ transformed state variable matrix, Eq. (74) and $(2N \times 1)$ truncated form

$\bar{y}^s, \bar{y}$ = $(6N_s \times 1)$ matrix of deformational nodal displacements for finite element s , and generic form

Greek Symbols

$\bar{\alpha}$ = coefficient of thermal expansion

β^j_α = rotation of nodal body j for axis $\underline{a}_\alpha$ due to variational nodal deformation) $(\alpha = 1,2,3)$

$\underline{\beta}^j, \beta^j$ = $\beta^j_1 \underline{a}_1 + \beta^j_2 \underline{a}_2 + \beta^j_3 \underline{a}_3$, and (3×1) matrix in basis $\{\underline{a}\}$

Γ = $(6N \times 1)$ matrix in Eq. (5)

$\varepsilon_{\alpha\gamma\theta}$ = epsilon symbol of tensor analysis (value $+1$, -1 or 0)

$\bar{\varepsilon}_{\alpha\gamma}$ = strain element, basis $\{\underline{e}\}$

$\bar{\varepsilon}$ = (6×1) strain matrix due to variations from steady-state deformation, Eq. (12)

$\bar{\varepsilon}'$ = steady-state strain matrix (6×1)

$\bar{\varepsilon}_\tau$ = (6×1) strain matrix due to deviations from steady-state temperature

ξ, η, ζ = Cartesian coordinates corresponding to $\underline{e}_1, \underline{e}_2, \underline{e}_3$ and origin fixed in element under steady-state deformation

Θ = (3×3) direction cosine matrix in $\{\underline{b}\} = \Theta\{\underline{i}\}$

$\bar{\kappa}^s, \bar{\kappa}$ = $(6N \times 6N)$ centripetal stiffness matrix for finite elements, and generic $\bar{\kappa}^s$

$\Lambda, \bar{\Lambda}$ = $(12N \times 12N)$ diagonal matrix of eigenvalues of B , and truncated $(2N \quad 2N)$ form

μ = mass density of finite element

ν = Poisson's ratio

$\underline{\rho}^s, \bar{\rho}^s, \rho, \bar{\rho}$ = position vector and (3×1) matrix in $\{\underline{e}^s\}$ basis to field point of element s in steady-state from CM^s_s; generic $\underline{\rho}^s$ and $\bar{\rho}^s$

$\bar{\sigma}'$ = steady-state stress (e.g. due to spin)

$\bar{\sigma}_{\alpha\gamma}$ = stress due to deviation from steady state deformation, $\{\underline{e}\}$ basis

$\bar{\sigma}$ = (6×1) stress matrix, Eqs. (16), (17)

$\bar{\sigma}_{\tau}$ = (6×1) stress matrix accomodating thermal strains

τ = variation from steady-state temperature

Φ = $(12N \times 12N)$ transformation matrix of eigenvectors, Eq. (74)

Φ' = $(12N \times 12N)$ matrix of adjoint eigenvectors of Eq. (71), see Eq. (72)

$\underline{\omega}^j$ = inertial angular velocity of nodal body j

$\underline{\omega}^a, \omega^a$ = inertial angular velocity vector and (3×1) matrix in $\{\underline{a}\}$ basis

$\underline{\omega}, \omega$ = inertial angular velocity vector and (3×1) matrix in basis $\{\underline{b}\}$

Ω = nominal value of ω^a, with elements $\Omega_1, \Omega_2, \Omega_3$

Operational Symbols

$(\ )^T$ = indicates matrix transposition

$(\ )^{\sim}$ or $(\tilde{\ })$ = indicates formation of (3×3) skew symmetric matrix from (3×1) matrix, as in Eq. (30)

$\frac{^f d}{dt}(\underline{V})$ = time derivative of arbitrary vector $\underline{V}$ in reference frame f

$(\dot{\ })$ = time derivative of scalar or matrix

$(\ )^*$ = virtual quantity (stress, displacement, etc.); also conjugate

$(\ )^{-1}$ = matrix inverse

$\triangleq$ = means equality by definition

Repeated lower case Greek indices indicate summation over range 1,2,3.

10. Many-body formalism

In principle, there has been no problem in handling the dynamics of systems of rigid bodies since the time of Euler. Those who might have needed to describe multi-body systems could use the free-body approach and write rotational and translational equations separately for each body, with proper attention to the interaction forces and torques among them, or they could turn to the alternative general formalism of Lagrange. Actually, there was little need for them to do either until recently, for there was not much that could be done with the equations after they were written except under very special circumstances, such as in problems that could be linearized in the dependent variables.

By the early 1960's, the aerospace application areas requiring multi-body dynamic modeling were in clear view, and so were feasible methods of numerical attack on such systems. Some general vector equations for systems of bodies had been developed following the Eulerian path, and within many organizations people were deriving and rederiving equations -- using both Eulerian and Lagrangian methods -- for systems of immediate parochial interest. Two deficiencies in the state of matters were apparent, however. First, the vector formulations required some manipulation and "translation" to be worked into scalar equations suitable for computer simulation. This process often was not very automatic and required some care in interpretation of quantities, even the participation of the original author of the equations for clarification, before one could be sure he had gotten final correct equations phrased in terms of constants and variables that really were those respectively known for and desired for the system at hand. Second, the final results were usually difficult, or at least messy, to generalize to systems other than the rather specific one in mind at the outset. A change in system configuration any more drastic than could be subsumed by simple adjustment of parameters, and such changes are typical rather than exceptional in the preliminary design of aerospace vehicles, would necessitate a good deal of extra work in the modification of the equations and in the programming of their digital simulation.

I am sure that these deficiencies became apparent to a number of people at about the same time, for several formal methods of developing multi-body dynamical equations from Eulerian principles were constructed subsequently, with the needs of simulation more or less explicitly in mind. Some have been published, others undoubtedly are used as working tools or lie buried in the cabinets of specific organizations.

My own response to the felt need, on which much of the present section is based, is described in a historical note appended to the section, which also includes a broad, interpretative survey of related developments. At this point it is enough to say that the underlying philosophy in the construction of the formalism that follows has been that it be structured with the needs of digital simulation in mind at the outset. It is algorithmic, the nature of the manipulative operations independent of the number of bodies and the precise way they are hooked together in specific applications, at least for broad classes of systems of bodies. The preparation of a problem for digital simulation involves minimal manipulative operations: as much as possible of the burden of routine calculation is shifted to the computer.

In mid-1964 I was closely involved with Dr. Gabriel Margulics on various problems of dynamical modeling. In one of our working sessions we agreed on the premise that a new Eulerian-type of dynamical formalism was a worthy and attainable goal, and on the general characteristics such a formalism should have. Afterwards, we each developed our own approach: his, with Robert Hooker, more geometrical; mine more algebraic. Shortly thereafter, Jens Wittenburg contributed a number of improvements to mine, and our joint work was published in 1966. The following material is largely based on that work and on its subsequent evolution in both our hands.

We consider a completely general system of bodies, each of which is in general, a non-rigid gyrostat. Imagine these being grouped into clusters of bodies, each cluster being defined as follows: If the bodies in it are put into correspondence with the vertices of a graph, as illustrated in Fig 1, the edges of the graph being drawn to represent kinematical constraints (e.g. hinge points) or force interactions among bodies, then the graph must:

1. have a cyclomatic number exceeding 1;
2. contain no isolated vertex type shown dashed in Fig. 1. (Formally, we can say there is no edge whose cutting separates the graph into two parts.)

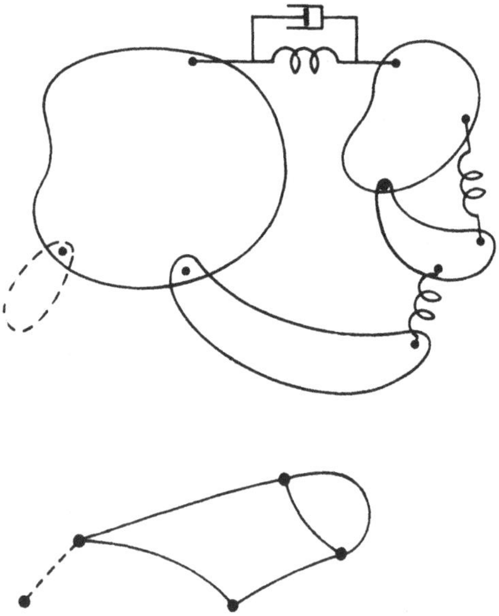

Fig. 1 A Cluster of Bodies and its Graph

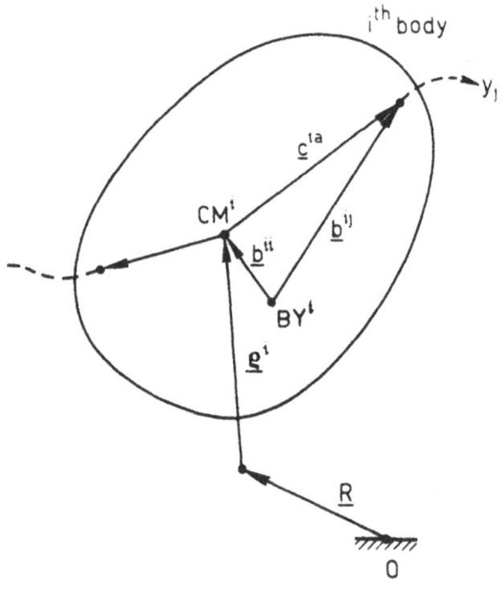

Fig. 2 The ith Cluster

An important practical case occurs when the role of a cluster is played by an elastic structure constituted of a number of simple elastic elements (beams, facets, etc.). Finally, imagine the clusters to be joined together in such a way that their graph is a tree, when each cluster is associated with a vertex and each interconnection between clusters is represented by an edge of the graph.

Every system of bodies can be described this way. Consider an arbitrary cluster, say the ith, whose center of mass CM^i is described by the vector $\underline{R} + \underline{\underline{\rho}}^i$ relative to an inertially fixed point O. The vector $\underline{R}$ alone locates the center of mass of the complete system (See Fig. 2). This cluster is attached to others at the "attachment points," called "hinge points" in the original formalism where they were always fixed in the body. The ath attachment point of the system, if it on the ith cluster, is located with respect to CM^i by vector $\underline{c}^{ia}$.

We shall find later that a very basic role is played by "augmented bodies" and their barycenters. Let us consider a fictitious cluster exactly like the ith cluster except that at each attachment point is placed a point mass equal in magnitude to the total mass of all clusters in that part of the system reached through that attachment point. This is the augmented cluster or body. The center of mass of the augmented cluster is called the barycenter of the real cluster. The vector from the barycenter of mass is denoted $\underline{b}^{ii}$. The vector $\underline{b}^{ij}$ is from the barycenter to the attachment point through which cluster j is reached. Later we shall get expressions for these vectors in terms of the $\underline{c}^{ia}$.

The Newtonian dynamical equation valid for any material system or subsystem gives us the translational dynamical equation for the cluster. Suppose that $\underline{F}^{i}$ is the total force from external effects on the ith cluster, and $\underline{\mathfrak{F}}^{a}$ is a force applied on it or by it through the ath attachment point. The equation

$$(1) \qquad m^{i}\frac{d^{2}}{dt^{2}}(\underline{R}+\underline{\rho}^{i}) = \underline{F}^{i} + \sum_{a} S^{ia}\underline{\mathfrak{F}}^{a}$$

is the translational equation. The coefficients S^{ia} are discussed in the next lecture: suffice it to say here that they are ± 1 or 0, chosen in such a way that the term $\sum_{a} S^{ia}\underline{\mathfrak{F}}^{a}$ does represent the effect on the ith cluster of all inter-cluster forces.

The Euler dynamical equations give us a corresponding set of rotational equations. Let $\mathbf{I}^{i}$ be the inertia dyadic of the cluster about its center of mass, $\underline{\omega}^{i}$ be the angular velocity with respect to inertial space of a body frame X^{i} associated with the ith cluster, $\underline{h}^{i}$ be the relative angular momentum of the cluster with respect to this frame, so

$$(2) \qquad \underline{H}^{i} = \mathbf{I}^{i}\cdot\underline{\omega}^{i} + \underline{h}^{i}$$

is the total angular momentum of the cluster about its center of mass. Further, let L^{i} be the total torque on the cluster about its center of mass from external effects, and $\underline{\mathfrak{L}}^{a}$ be the torque on or by it at the ath attachment point. The equations are

$$\frac{d\underline{H}^{i}}{dt} = \underline{L}^{i} + \sum_{a} S^{ia}(\underline{\mathfrak{L}}^{ia} + \underline{c}^{ia} \times \underline{\mathfrak{F}}^{a}) . \tag{3}$$

Again the coefficients S^{ia} properly insert the interaction torques.

Our task will be to begin with Eqs. 1-3 and convert them to the simplest matrix form.

We are primarily interested in rotation, though we may encounter relative translation between clusters as well. The translational equations are used, first of all, to eliminate the $\underline{\mathfrak{F}}^{a}$ terms from Eq. 3. Thereafter, as many of them are retained as are needed to express relative translation, but the rest are discarded. The remainder of the theory concerns the reduction of Eq. 3.

11. Many-body systems: graph-theoretic approach

In the previous lecture the system was described for which now a dynamical formalism is to be developed. The system was said to consist of an arbitrary number of rigid bodies. It has a topological tree-structure. Its hinges of one, two or three degrees of freedom of relative rotational motion contain at least one point which is a fixed point on either one of the two bodies connected by the respective hinge. This point was called the hinge point. A tree-structure of n bodies has $n-1$ hinge points. The dynamical formalism to be developed must combine the following advantages:

a) it must be exact, i.e. no approximations are accepted

b) it must lead to equations the various terms of which are simple to interprete physically

c) it must be easily applicable to mechanical problems of different nature

d) it must be formulated in such a way that it can be easily put on a digital computer

In all problems of mechanics the choice of variables for the description of the system is one of the crucial points on which success of failure depends. It was pointed out already that it seems to be most promising to separate the dynamical from the kinematical equations of motion by selecting as variables angular velocity components resolved in body-fixed coordinate systems. As important as the choice of variables is a skilfull choice of mathematical tools used for the description of the system behaviour. It was said already that it is practical to start out with equations of motion written in vector form. Since, ultimately, we want to arrive at a set of scalar equations the use of matrices seems to be suited best. Thus, methods must be found which allow the transition form vector to matrix equations. A first step in this direction will be demonstrated in this lecture. One of the most important problems to be solved, however, will be the one posed by the arbitraryness of the topological structure of the system. A system of two bodies does not seem to present unsurmountable difficulties since two bodies will always form a chain. The same applies even to the case of three rigid bodies. If, however, four or more bodies are interconnected, then there are more and more apparently entirely different systems. Thus, five bodies can be arranged in the form of a chain without branching points. They can also be arranged in the form of a star with one central body and four bodies attached to this central body. Or, finally, they can form a system of the kind shown in Fig. 1. Obviously, the number of topologically different arrangements of n bodies (*) is rapidly increasing with increasing n. The task of formulating a system of equations which is equally valid for all values of the number n and for all possible different arrangements calls for a mathematical tool which is specifically designed for the description of topological structures. Such a tool is found in graph theory. Graph theory, originally created by Euler as a branch of pure mathematics has become important in recent years in many fields of application. Electrical networks, railroad and car traffic networks, distribution channels for products, decision sequences in logical networks, flow charts of computer programs, job sequences in industrial projects (CPM, PERT etc.) etc. are typical examples. They all have in common that points (identifi-

(*) The number was calculated as a function of n by Cayley.

ed as road intersections for example) and lines connecting these points (identified as roads for example) are forming a network. When such a system is stripped of its physical interpretation a structure is left which is called graph and which consists of a number of vertices (the name for the points mentioned above) connected by arcs. The system of n interconnected bodies which we are considering can be described by such a graph, as well. The bodies are symbolized by the vertices and the connecting hinge points by the arcs. The graph of Fig. 2 is the graph belonging to the system of Fig. 1. To the author's knowledge this is the first application of graph theoretical concepts to rational mechanics. As will be seen, it is a very successful approach. Among the examples of applications of graph theory given above there are some in which the information that two vertices are connected by an arc must be accompanied by a statement of a sense of direction. For example, in a network of product distribution channels the arcs require an indication of the direction of transportation. This can be done by arrows along the arcs. Such graphs with directions assigned to the arcs are called directed graphs. In our dynamical system it is necessary to define arc directions for the following reason. Two bodies of the system which are directly interconnected by a hinge exert hinge reaction forces on each other. The two forces are equal in magnitude and opposite in direction. If we assign one reaction force to the hinge point then we must specify on which of the two bodies it is acting with negative sign. This is most easily done by assigning a sense of direction to the arc symbolizing the hinge point and by stating that the hinge reaction force is meant to act with positive sign on the body (i.e. the vertex) where the (arrow on the) arc starts. Fig. 3 shows the directed graph for the system of Fig. 1. Together with fixing arc directions the vertices and arcs have been numbered. This can be done in an arbitrary fashion. It turns out to be advantageous later to assign the number one to a vertex on a branch end as shown in Fig. 3. The arc directions can be assigned in an arbitrary way, as well. It is advantageous, however, to point the arrows in such a way that they are all directed towards vertex number one. The graph of a system with tree-structure always has branch ends and the arc directions are unambiguously defined once vertex number one has been chosen. For systems with closed loops this is not the case. As we said, however, numbers and arc directions could have been chosen in any other, way, as well. The restriction to systems with tree-structure has another reason which will become apparent later. Subsequently, the vertices and arcs will be referred to as $s_i (i=1,\ldots,n)$ and $u_a (a=1,\ldots,n-1)$, respectively.

On the directed graph of Fig. 3 a so-called incidence matrix $\underline{S}$ is defined as follows. It has n rows and $n-1$ columns corresponding to the number of vertices and arcs, respectively. Its elements S_{ia} (first index = row index) are

$$(1) \qquad S_{ia} = \begin{cases} +1 & \text{if arc} \quad \text{starts at vertex } s_i \\ -1 & \text{" " " ends " " "} \\ 0 & \text{otherwise} . \end{cases}$$

For the graph of Fig. 3 $\underline{S}$ has the form

$$\underline{S} = \begin{bmatrix} . & . & . & -1 \\ . & -1 & +1 & . \\ +1 & . & . & . \\ -1 & . & -1 & +1 \\ . & +1 & . & . \end{bmatrix} .$$

From (1) follows that each column of $\underline{S}$ contains exactly one element +1 and one element −1. Each row contains as many non-zero elements as there are arcs connected with the vertex correspondig to the row index. This property allows the following elegant formulation. Let $[\vec{X}_a]$ be a column matrix made up of the $n-1$ hinge reaction forces $\vec{X}_1,\ldots,\vec{X}_{n-1}$. Then the expression $\underline{S}[\vec{X}_a]$ is a column matrix of n elements each one of which is the sum of all hinge reaction forces acting on the body represented by the index of the element. Body number 4 in Fig. 3, for instance, is subject to the resultant reaction force $-\vec{X}_1 - \vec{X}_2 + \vec{X}_4$. The distribution of elements +1 , −1 and zero in the matrix S automatically causes each hinge reaction force to appear in exactly two elements of the column matrix $\underline{S}[\vec{X}_a]$ and with opposite algebraic sign. This property of the incidence matrix $\underline{S}$ would justify already the use of graph theory. However, there exists another matrix on the graph of Fig. 3 which is equally important for the subsequent theory. This matrix, which will be called $\underline{T}$ is an $[(n-1)\times n]$ -matrix with elements

$$T_{ai} = \begin{cases} +1 & \text{if arc } u_a \text{ is on the direct path between the vertices } s_i \text{ and } s_1 \\ 0 & \text{otherwise .} \end{cases} \qquad (2)$$

The direct path between the vertices s_i and s_1 is the minimal set of vertices and arcs which has to be passed when travelling from s_i to s_1 . It includes s_i and s_1 . For the graph of Fig. 3 the matrix is

$$\underline{T} = \begin{bmatrix} . & . & 1 & . & . \\ . & . & . & . & 1 \\ . & 1 & . & . & 1 \\ . & 1 & 1 & 1 & 1 \end{bmatrix} .$$

The above definition of $\underline{T}$ requires to have a graph with tree-structure. Otherwise the direct path between two vertices cannot be defined. This is a mathematical reason for our restriction to tree-structures. The underlying physical reason will become apparent soon. It should be recognized that the first column of $\underline{T}$ always contains only zeros. The matrices $\underline{S}$ and $\underline{T}$ which, both describe the structure of the system obey the fundamental

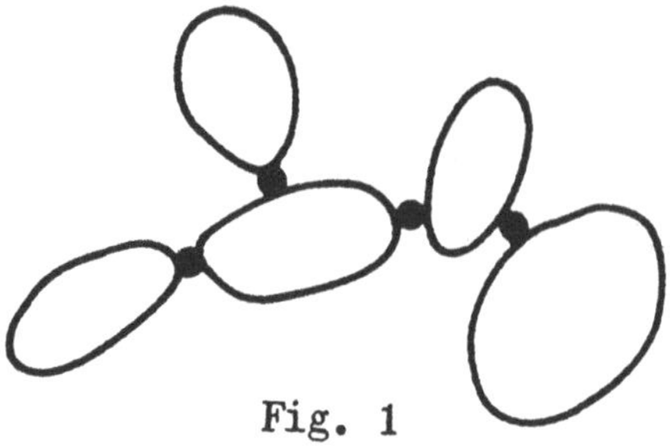

Fig. 1

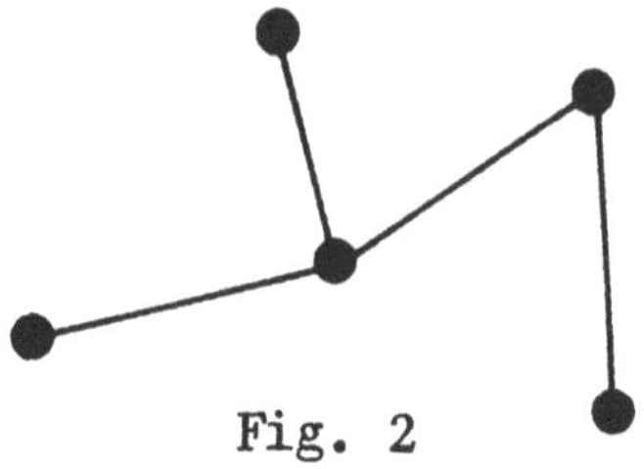

Fig. 2

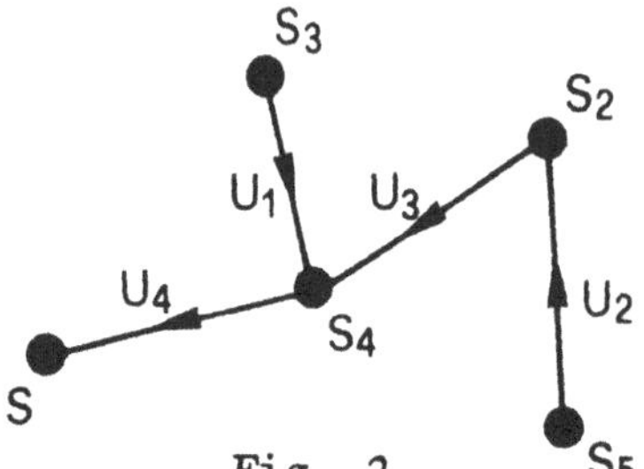

Fig. 3

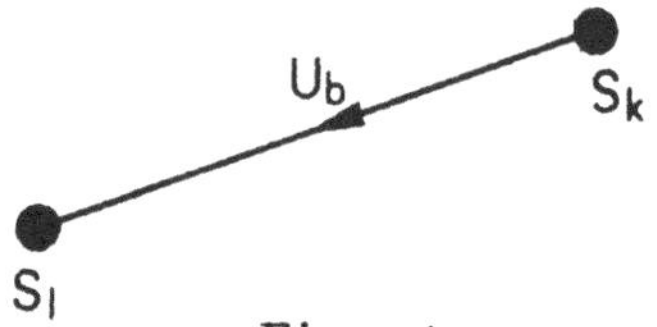

Fig. 4

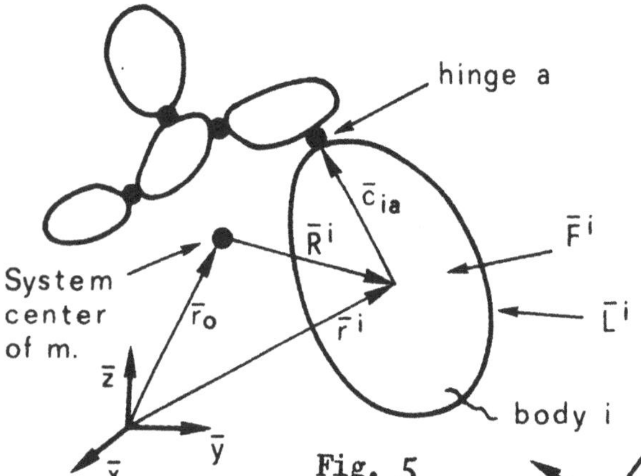

Fig. 5

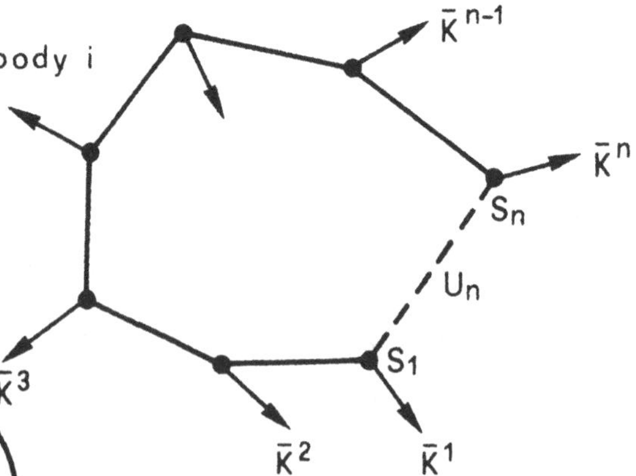

Fig. 6

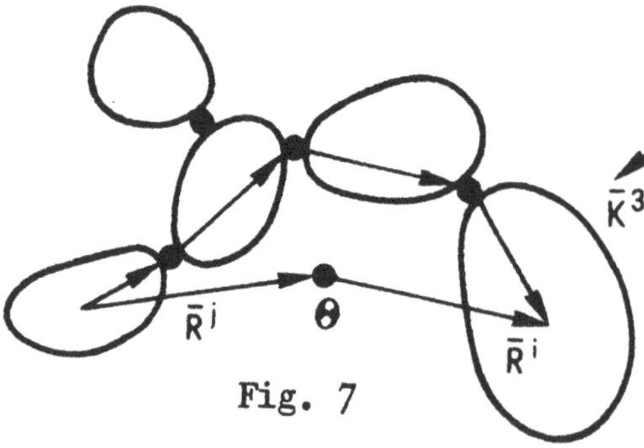

Fig. 7

Theorem 1: The product $\underline{T}\,\underline{S}$ is a unit matrix.

Proof: Since each column of $\underline{S}$ has exactly two non-zero elements the elements of $\underline{T}\underline{S}$ are $(\underline{T}\,\underline{S})_{ab} = T_{ak} - T_{a\ell}$ $(a,b = 1,\ldots,n-1)$ with vertices s_k and s_ℓ arranged as shown in Fig. 4. If $a=b$ then $T_{ak}=1$, $T_{a\ell}=0$. Hence $(\underline{T}\,\underline{S})_{aa}=1$. If $a \neq b$ then either $T_{ak} = T_{a\ell} = 1$ or $T_{ak} = T_{a\ell} = 0$ and consequently $(\underline{T}\,\underline{S})_{ab} = 0$. End of proof.

This theorem enables one to eliminate the hinge reaction forces from the equations of motion as will be shown now. The translational motion of a representative body number i can be written in the form

$$m^i \ddot{\vec{r}}^{\,i} = \vec{F}^{\,i} + \sum_{a=1}^{n-1} S_{ia} \vec{X}_a\,, \qquad i = 1,\ldots,n \tag{3}$$

where $\vec{r}^{\,i}$ is the radius vector of the body i center of mass in an inertial coordinate system $\vec{x},\vec{y},\vec{z}$ (Fig. 5) and $\vec{F}^{\,i}$ is the resultant external force. The sum on the right side is the i-th element of the column matrix $\underline{S}[\vec{X}_a]$ which was shown to be the sum of all hinge reaction forces on body i. It is advantageous to rewrite (3) in terms of the radius vectors $\vec{R}^{\,i}$ directed from the system center of mass to the body i center of mass. Then the over-all translation of the system is described by the single expression

$$M\ddot{\vec{r}}_0 = \sum_{j=1}^{n_1} \vec{F}^{\,j} \tag{4}$$

$(M = \sum_{i=1}^{n} m^i)$ whereas the equations for the various bodies are formulated in terms of motion relative to the moving system center of mass. From

$$\vec{r}^{\,i} = \vec{r}_0 + \vec{R}^{\,i}$$

follows with (4)

$$m^i \ddot{\vec{r}}^{\,i} = \frac{m^i}{M} \sum_{j=1}^{n} \vec{F}^j + m^i \ddot{\vec{R}}^{\,i} .$$

Substitution into (3) yields

$$(5) \qquad m^i \ddot{\vec{R}}^{\,i} = \sum_{j=1}^{n} \mu_{ij} \vec{F}^j + \sum_{a=1}^{n-1} S_{ia} \vec{X}_a , \qquad i = 1,\dots,n$$

with $\mu_{ij} = \delta_{ij} - m^i/M \ (i,j = 1,\dots,n)$ and with the Kronecker symbol δ_{ij}. All n vectorial equations (5) can now be combined in the single matrix equation

$$(6) \qquad \underline{m}\left[\ddot{\vec{R}}^{\,i}\right] = \underline{\mu}\left[\vec{F}^{\,i}\right] + \underline{S}\left[\vec{X}_a\right]$$

Here, m is the $(n \times n)$-diagonal matrix of the masses $m^1,\dots,m^n$, $\underline{\mu}$ the $(n \times n)$-matrix with the elements μ_{ij} and $\left[\ddot{\vec{R}}^{\,i}\right]$ and $\left[\vec{F}^{\,i}\right]$ are column matrices of n vectorial quantities each similar to the term $\left[\vec{X}_a\right]$ which was introduced earlier already. The matrix $\underline{\mu}$ is singular since the sum of all rows is a row containing only zeros. Apart from the matrix $\underline{S}$ Eq. (6) is unconventional because of the use of matrices the elements of which are vectors. By multiplying (6) from the left side with $\underline{T}$ we arrive, because of The orem 1, at an explicite expression for the hinge reaction forces:

$$[\vec{X}_a] = \underline{T}\left(\underline{m}[\ddot{\vec{R}}^i] - \underline{\mu}[\vec{F}^i]\right) . \qquad (7)$$

At this point the physical reason becomes apparent which necessitates the restriction to systems with tree-structure. Eq. (6) expresses a state of static equilibrium for each individual body between the resultant forces $\vec{K}^i = m^i\ddot{\vec{R}}^i - \sum_{j=1}^{n}\mu_{ij}\vec{F}^j \ (i = 1,\dots,n)$ on one hand and hinge reaction forces on the other. It is easily seen that $\sum_{i=1}^{n}\vec{K}^i = 0$. In Fig. 6 a chain, i.e. a tree-structured system of an unspecified number n of vertices is schematically shown with the forces $\vec{K}^i$ applied to the vertices. It is obvious that forces $\vec{X}_1,\dots,\vec{X}_{n-1}$ can be determined which equilibrate arbitrary loading forces $\vec{K}^i$ (provided that $\sum_{i=1}^{n}\vec{K}^i = 0$).
One need only start at one end of the chain and determine one hinge reaction force after the other. Now, suppose that the vertices number 1 and number n are coupled be an arc u_n shown as a dashed line. The loading $\vec{K}^1,\dots,\vec{K}^n$ being the same as before equilibrium exists if $\vec{X}_1,\dots,\vec{X}_{n-1}$ are given the previously calculated magnitudes while $\vec{X}_n$ is set equal to zero. However, one might equally well set equal to zero any other hinge reaction force and then uniquely determine the remaining ones. The system is statically indeterminate. This situation is not altered if the closed loopsystem is given a more general form by connecting additional branches of verticesand arcs to some of the vertices in Fig. 6. The statical indeterminateness finds its mathematical

expression in the fact that for systems with closed loops (n vertices and m arcs with $m > n-1$) the $(n \times m)$-incidence matrix $\underline{S}$ has no left inverse $\underline{T}$ such that $\underline{T}\,\underline{S}$ is a unit matrix. This is a generalresult of graph theory.

Before Eq. (7) can be used further the dynamical equations of rotational motion must be formulated in a way comparable with (6). For body i alone the equation holds

$$(8) \qquad \dot{\vec{H}}^i = \vec{L}^i + \sum_{a=1}^{n-1} S_{ia}\vec{Y}_a + \sum_{a=1}^{n-1} S_{ia}\vec{c}_{ia} \times \vec{X}_a \,, \qquad i = 1,\dots,n \,.$$

By $\vec{H}^i$, $\vec{L}^i$ and $\vec{Y}_a$ the absolute angular momentum, the resultant external torque and the hinge reaction torque in hinge a are denoted, respectively. One word about $\vec{Y}_a$: Hinge reaction torques can be caused by springs and dash pots resisting relative rotational motions of coupled bodies or they can be due to the fact that a hinge gives less than three degrees of freedom of relative rotational motion to the coupled bodies. This happens in the case of an axis joint, for example. At this point of the development hinges will be regarded as if they were frictionless, spherical hinges. All torques caused by the fact that, in reality, they are not frictionless, spherical hinges are combined in the terms $\vec{Y}_a$. More details about this will be said in one of the next lectures. In Eq. (8) the vector $\vec{c}_{ia}$ is the lever on which the hinge reaction force $\vec{X}_a$ produces a torque about the body i center of mass (Fig. 5).

The sums over all hinges $a=1,\dots,n-1$ are equivalent to similar expressions in (3) and (5). The concentration of all n separate equations (8) in a single matrix equation comparable with the transition from (5) to (6) is straight forward except for the last term on the right side which requires special attention. As a first step a matrix $\underline{\vec{C}}^*$ can be constructed which has the same dimension as the incidence matrix $\underline{S}$, i.e. it is of size $n\times(n-1)$. Its elements are, however, not the scalars S_{ia} but the vectors $S_{ia}\vec{c}_{ia}$. The vector character of the elements is indicated by the arrow on the symbol $\underline{\vec{C}}^*$. Then the expression $\left[\sum_{a=1}^{n-1} S_{ia}\vec{c}_{ia}\times\vec{X}_a\right]$ which is the column matrix of all n sums can be given the form $\underline{\vec{C}}^*\otimes[\vec{X}_a]$ if the multiplication symbol $\otimes$ is defined as follows. Let $\underline{\vec{A}}$ be an $(m\times n)$ -matrix with vector elements b_{ij}. Then, by the product $\underline{\vec{C}}=\underline{\vec{A}}\otimes\underline{\vec{B}}$ an $(m\times p)$-matrix is meant whose elements are vectors C_{ij} calculated from the equation

$$\vec{c}_{ij} = \sum_{k=1}^{n}\vec{a}_{ik}\times\vec{c}_{kj}\,, \quad i=1,\dots,m\,, \quad j=1,\dots,p\,.$$

The symbol $\otimes$, thus, indicates an algebraic manipulation which follows the usual rules of matrix multiplication except that any product of two elements is to be understood as a vector cross product. It is easily verified that with this definition the expression $\underline{\vec{C}}^*\otimes[\vec{X}_a]$ is identical with the column matrix $\left[\sum_{a=1}^{n-1} S_{ia}\vec{c}_{ia}\times\vec{X}_a\right]$. This is a convenient place to define still another kind of product which will be needed later. With $\underline{\vec{A}}$ and $\underline{\vec{B}}$ being the same matrices as above the expression $\vec{\underline{A}} \otimes \vec{\underline{B}}$ (referred to as scalar matrix product) is defined as the scalar matrix C of size $(m \times p)$ with elements

$$c_{ij} = \sum_{k=1}^{n} \vec{a}_{ik} \cdot \vec{b}_{kj} \,, \quad i = 1,\dots,m \,, \quad j = 1,\dots,p \,.$$

Now we return to Eq. (8). All n separate equations can be concentrated in the single matrix equation

$$(9) \qquad [\dot{\vec{H}}^i] = [\vec{L}^i] + \underline{S}[\vec{Y}_a] + \underline{\vec{C}}^* \otimes [\vec{X}_a]$$

where column matrices $[\dot{\vec{H}}^i]$, $[\dot{\vec{L}}^i]$ and $[\vec{Y}_a]$ have been defined which combine the n vectors $\dot{\vec{H}}$ and $[\dot{\vec{L}}^i]$ and the $n-1$ vectors $\vec{Y}_a$ in the same way as the $n-1$ vectors $\vec{X}_a$ are combined in the column matrix $[\vec{X}_a]$. With the help of the matrix $\underline{T}$ we had succeeded in solving (6) for $[\vec{X}_a]$ explicitely. Substitution of (7) into (9) yields the equation

$$(10) \qquad [\dot{\vec{H}}^i] = [\vec{L}^i] + \underline{S}[\vec{Y}_a] + \underline{\vec{C}}^* \otimes \underline{T}\left(\underline{m}[\ddot{\vec{R}}^i] - \mu[\vec{F}^i]\right) .$$

The set of equations (4), (6) and (9) or which is the same the set of equations (4), (7) and (10) must be accompanied by geometric compatibility conditions expressing the fact that every hinge point is a fixed point on two bodies simultaneously. The formulation of these conditions will be shown in detail in the next lecture. At this point, a few words indicating the essential ideas will help to appreciate the achievement gained with Eq. (10). The difference of any two vectors $\vec{R}^i - \vec{R}^j$ can be expressed as a sum of body-fixed vectors as is indicated in Fig. 7. The fact that the point $\mathfrak{F}$ is the system center of mass provides one additional linear relationship between the vectors $\vec{R}^i (i = 1,\dots,n)$. This equation enables one to express the vectors $\vec{R}^i (i = 1,\dots,n)$ themselves as sums of body-fixed vectors. The term $[\ddot{\vec{R}}^i]$ in (10), consequently, can be expressed in terms of body-fixed vectors, of angular velocities and angular accelerations. Eq. (10), thus, turns out to be an equation of rotational motions only.

As a final step of preparation for the introduction of new basic relationships in the next lecture a separation of unit vectors and scalar vector components will be carried out. For this purpose, on each body of the system a body-fixed cartesian coordinate system with unit vectors $\vec{e}_1^i, \vec{e}_2^i, \vec{e}_3^i (i = 1,\dots,n)$ is defined. The unit vectors on body i are combined in a column matrix $\underline{\vec{e}} = [\vec{e}_1^i\ \vec{e}_2^i\ \vec{e}_3^i]^T$ (the superscript T denotes transposition). The vector $\vec{c}_{ia}$ which is fixed on body i has, in the coordinate system $\underline{\vec{e}}^i$, the constant components $c_{ia1,2,3}$ which form a column matrix $\underline{c}_{ia} = [c_{ia1}\ c_{ia2}\ c_{ia3}]^T$. For the above quantities the equation holds $\vec{c}_{ia} = \underline{\vec{e}}^{iT} \underline{c}_{ia}$. This separation of unit vectors and vector components can be applied to any vector, for instance to the vector $\vec{L}^i$. The result is a relationship $\vec{L}^i = \underline{\vec{e}}^{iT} \underline{L}^i$ where $\underline{L}^i$ is the column matrix of the components of $\vec{L}^i$ resolved in $\underline{\vec{e}}^i$. Of course, $\underline{L}^i$ is a time-changing quantity whereas $\underline{c}_{ia}$ is a constant system parameter. If a $(3n \times n)$-matrix $\underline{\vec{e}}$ is introduced which is partitioned into the (3×1)-column matrices $\underline{\vec{e}}_{ij} = \delta_{ij}\underline{\vec{e}}^{i}(i,j = 1, 2,\ldots,n)$ then the column matrix $[\vec{L}^{i}]$ can be replaced by the product $\underline{\vec{e}}^{T}\underline{L}$ where $\underline{L}$ is a column matrix of $3n$ elements each submatrix of three elements being the component representation of a vector $\vec{L}^{i}$ resolved into the respective body-fixed coordinate system $\underline{\vec{e}}^{i}$. The same procedure can be applied to the terms $[\vec{H}^{i}]$ and $[\vec{F}^{i}]$ yielding the expressions $\frac{d}{dt}(\underline{\vec{e}}^{T}\underline{H})$ and $\underline{\vec{e}}^{T}\underline{F}$. Finally, also the matrix $\underline{\vec{C}}^{*}$ can be split up yielding the expression $\underline{\vec{e}}^{T}\underline{C}$ where $\underline{C}$ is a $[3n \times (n-1)]$ -matrix partitioned into the (3×1)-submatrices

$$\underline{C}_{ia} = S_{ia}\underline{C}_{ia}, \quad i = 1,\ldots,n, \quad a = 1,\ldots,n-1.$$

The matrix $\underline{C}$, thus introduced, is a constant quantity specifying the location of the hinge points on the various bodies. It will be referred to as the connection matrix. With these changes Eq. (10) takes the form

$$\frac{d}{dt}(\underline{\vec{e}}^{T}\underline{H}) = \underline{\vec{e}}^{T}\underline{L} + \underline{S}[\vec{Y}_{a}] + \underline{\vec{e}}^{T}\underline{C}\otimes\underline{T}(\underline{m}[\ddot{\vec{R}}^{i}] - \underline{\mu}\,\underline{\vec{e}}^{T}\underline{F}).$$

12. Barycentric Vectors and Related Topics

The last equation of the previous lecture shows, with minor differences in notation, the form taken by Eq. 3 of lecture 10 when Eq. 1 of that lecture is solved for a and the result used to eliminate the attachment interaction forces from the rotational equations.

Repeating the equation without yet resolving the various vectors into their component matrices, it is

$$\frac{d}{dt}(\mathbf{I}^i \cdot \underline{\omega}^i + \underline{h}^i) = \underline{L}^i + \sum_a S^{ia}\{\underline{\mathcal{L}}^a + \underline{c}^{ia} \times \sum_i T^{ai}[m^i \ddot{\underline{\rho}}^i - \mu^{ij}\underline{F}^j]\} \ .$$

Our task is to simplify this equation. (In this lecture, T^{ai} is denoted S^{*ai})

We begin by recalling the definition of an <u>augmented i^{th} body</u>, defined as the ith together with point masses placed at all the hinge points on the ith body, the mass at any one of these points being chosen equal to the total mass represented by all bodies that include the hinge point on their direct circuits to v_i. The ith body barycenter is defined as the center of mass of the augmented i^{th} body. Denote by $\underline{b}^{ii}$ the vector to the ith body center of mass from the ith body barycenter. If the direct chain from body j to body i reaches the latter at arc a (arc a is attached to the ith body), then the vector to the hinge point corresponding to arc a from the ith body barycenter is denoted $\underline{b}^{ij}$ (See Fig. 1).

To relate the elements of $[\underline{C}]$ to the barycentric vectors $\underline{b}^{ii}$ and $\underline{b}^{ij}$ we consider the products $[\underline{d}]=[\underline{C}]S^*$ and $[d]\mu$, and seek their physical interpretations. We call the matrix d corresponding to $[\underline{d}]$ the path matrix of the system.

On the vector level we may write $\underline{d}^{ij} = \underline{c}^{ia} S^{*aj} = \sum_a \underline{c}^{ia} S^{ia} S^{*aj}$. Consider five exhaustive and mutually exclusive possibilities:

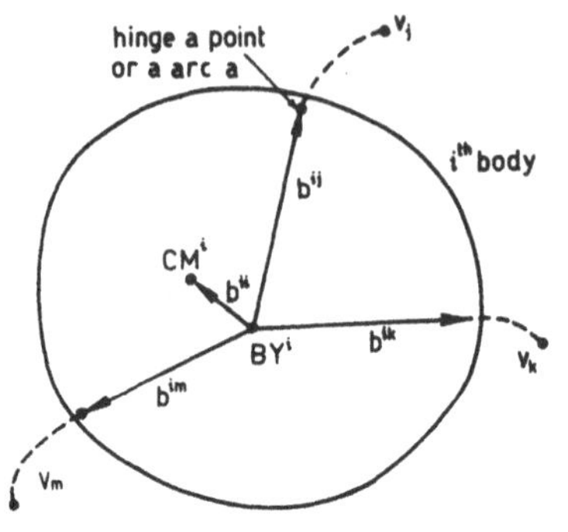

Fig. 1 Barycentric Vectors

1. $v_j = v_r$. $S^{*aj} = 0$ for all a, whence $\underline{d}^{ij} = 0$.

2. $v_j \neq v_r$, $v_i \not\leq v_j$. Denote by $\sum_j$ the set of arc indices on the direct chain between v_j and v_r. $S^{*aj} = 0$ if $a \notin \sum_j$. But v_j is not on this chain, so $S^{ia} = 0$ for $a \in \sum_j$. Again $\underline{d}^{ij} = 0$.

3. $v_j \neq v_r$, $v_i = v_j$. We know $S^{ai} = 0$ except for hinge points on the ith body. But also $S^{*aj} = 0$ unless $a \in \sum_i$. Hence the only surviving term is $\underline{c}^{ia} S^{ia} S^{*aj}$ with a on body i and a the specific hinge point on the direct chain from body i to v_r. If the ath arc is directed toward v_r, both S^{ia} and S^{*aj} are $+1$; if away, they are both negative. In any case, then, $\underline{d}^{ii} = \underline{c}^{ia}$ with the hinge point specified above.

4. $v_j \neq v_r$, $v_r \neq v_i < v_j$. The only surviving terms are those for which $a \in \sum_j$, but of these all are zero because of S^{ai} unless a designates a hinge point of the ith body. Thus two terms survive: at a_1, where the direct chain from v_j to

ward v_r enters v_i, and at a_2 where it leaves body i. If arcs a_1 and a_2 are both directed toward v_r, $\underline{d}^{ij} = \underline{c}^{ia_1}(-1)(+1) + \underline{c}^{ia_2}(+1)(+1) = \underline{c}^{ia_2} - \underline{c}^{ia_1}$. If arc a_1 is toward v_r and a_2 is away from v_r, $\underline{d}^{ij} = \underline{c}^{ia_1}(-1)(+1) + \underline{c}^{ia_2}(-1)(-1) = \underline{c}^{ia_2} - \underline{c}^{ia_1}$. (The same result follows for the other two possibilities.)

5. $v_j \neq v_r$, $v_r = v_i$. The only surviving term is the one for which a denotes the hinge point on v_r where the direct chain from v_j enters v_r. If that arc is directed away from v_r, $d^{ij} = \underline{c}^{ia}(-1)(+1) = -\underline{c}^{ia}$; if away, $\underline{d}^{ij} = \underline{c}^{ia}(-1)(+1) = -\underline{c}^{ia}$.

Collecting these results, we can say that

$$
d^{ij} = \begin{cases}
0 \text{ if } v_j = v_r \text{ or if } v_i \not\leq v_j \\
\text{vector from } v_i \text{ center of mass to first hinge point en} \\
\qquad \text{route from } v_i \text{ to } v_r, \text{ if } v_r \neq v_i = v_j \\
\text{vector across } v_i \text{ from hinge point where direct chain} \\
\qquad \text{from } v_j \text{ toward } v_r \text{ enters } v_i \text{ to hinge point where} \\
\qquad \text{it leaves } v_i, \text{ if } v_r \neq v_i < v_j \\
\text{vector to } v_r \text{ center of mass from last hinge point on} \\
\qquad \text{direct chain from } v_j \text{ to } v_r, \text{ if } v_r = v_i \neq v_j .
\end{cases} \tag{2}
$$

The results are illustrated in Fig. 2. The sub-matrix d^{ij}, of course, is simply the resolution of d^{ij} in the x^i-frame.

Having interpreted $\underline{d}^{ij}$, examine the product $[\underline{d}]\mu$ where μ was defined previously. We immediately encounter sums of the form $\sum_j \underline{d}^{ij} m^j$, so we begin by giving these a physical interpre-

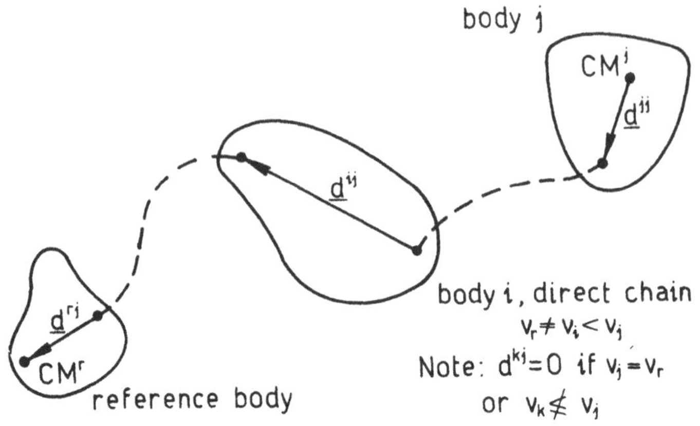

Fig. 2 Interpretation of $\underline{d}^{ij}$ Vectors

tation.

Refer to Fig. 3, where the ith body is shown with three typical hinge points, although any number can be considered. Assume that the direct chain from the ith body to the reference body originates at a_1. Denote by $\sum_1, \sum_2$, etc. the sets of bodies reached by direct chains from the ith

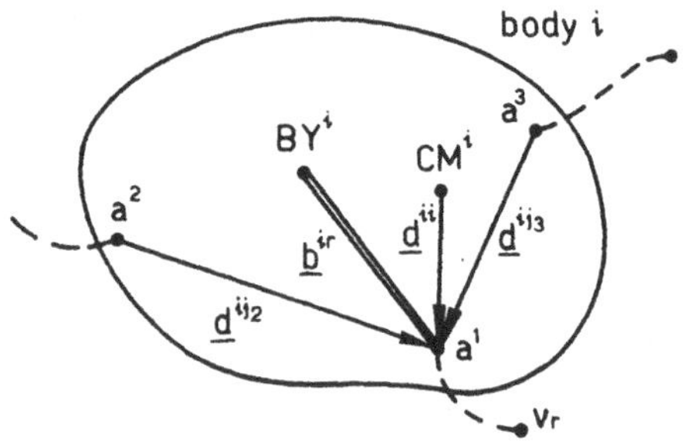

Fig. 3 The ith Body with d-vectors

body through the hinge points a_1, a_2, etc. The sum in question can be decomposed as

$$\sum \underline{d}^{ij} m^j = \underline{d}^{ii} m^i + \sum_{j \in \Sigma_1} \underline{d}^{ij} m^j + \sum_{j \in \Sigma_2} \underline{d}^{ij} m^j + \ldots .$$

For $j \in \Sigma_2$, $\underline{d}^{ij}$ is the vector $\underline{d}^{ij2}$ shown in the figure from hinge point a_2 to hinge point a_1, and similarly for $j \in \Sigma_3$ etc. For $j = i$, $\underline{d}^{ii}$ is the vector from the ith body center of mass to a_1. Let $m^{j2} = \sum_{j \in \Sigma_2} m^j$, etc. be the mass of all bodies of the system reached by direct chains from v_i through point a_2, and analogously for m^{j3} etc. Thus

$$\sum \underline{d}^{ij} m^j = m^i \underline{d}^{ii} + m^{j2} \underline{d}^{ij2} + m^{j3} \underline{d}^{ij3} + \ldots .$$

By definition given previously for the ith augmented body, we see that the negative of the right hand side is precisely the vector which locates the center of mass of the augmented body — hence the barycenter BY^i — with respect to the hinge point a_1. That is, for all (*) i,

$$\sum_j \underline{d}^{ij} m^j = m \underline{b}^{ir} . \tag{3}$$

We now use Eq. (3) to arrive at an interpretation of $[d]\mu$. The element of the product in the ith row j^{th} column is

---

(+) The argument has been given only for $v_i \neq v_r$. The same result is obtained for $v_i = v_r$.

$$(\underline{d}\mu)^{ij} = \underline{d}^{ij} - \frac{1}{m}\sum_j \underline{d}^{ij} m^j = \underline{d}^{ij} - \underline{b}^{ir}$$

Figure 4 shows the ith body, assumed not to be coincident with either the jth body or the reference body. From what we have already established about the meaning of $\underline{d}^{ij}$, it is evident that $\underline{b}^{ij} + \underline{d}^{ij} = \underline{b}^{ir}$. If we denote by $[\underline{b}]$ the array $[\underline{b}^{ij}]$, in accordance with the notational conventions laid down previously, we conclude (*) that

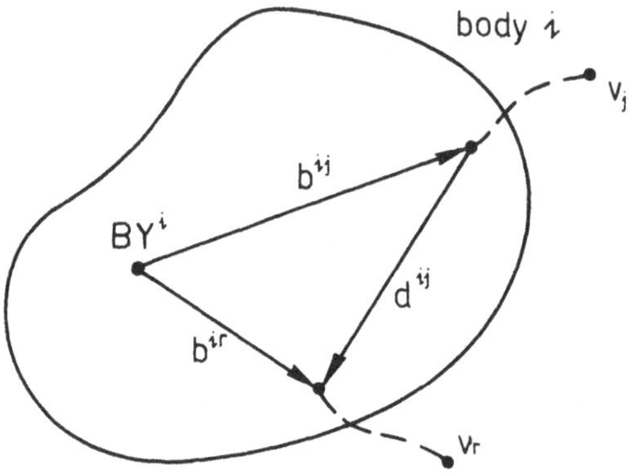

Fig. 4 Relationship between d^{ij} and Barycentric Vectors

(4) $$[\underline{d}]\mu = -[\underline{b}] .$$

We now have defined and discussed $[\underline{d}]$ and $[\underline{b}]$ and have shown that $[\underline{d}]\mu = [\underline{b}]$. The motivation for this sequence of development is not apparent, but is simply the fact that such combinations as $[\underline{d}]$ and $[\underline{d}]\mu$ arise in a fundamental way during the

(+) The conclusion is valid for $v_i = v_j$ and $v_i = v_r$ as well.

reduction of the equations. One more step completes the construction of the machinery we need to carry out that reduction. It is the expression in terms of $\underline{b}^{ij}$ of the vectors $\underline{\varrho}^{i}$ locating the center of mass of the ith body with respect to the center of mass of the system.

Recognize that $\sum_i \underline{d}^{ik}$ is the sum of the chain of vectors stretching from the center of mass of body k to that of the reference body, as shown in Fig. 5. Thus

$$\sum_i \underline{d}^{ik} = \underline{\varrho}^{r} - \underline{\varrho}^{k} . \tag{5}$$

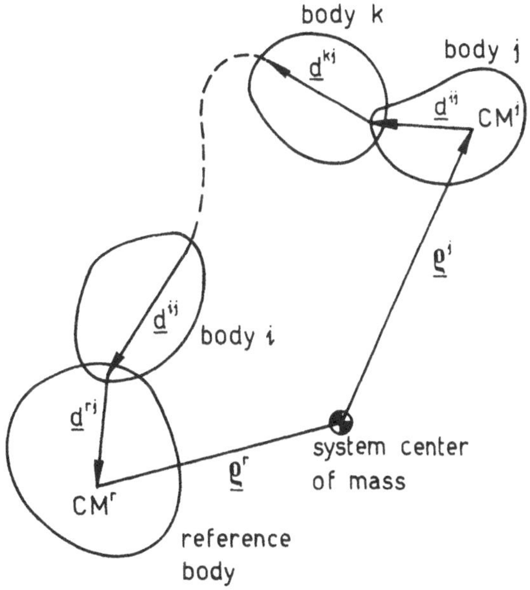

Fig. 5 Relationship between the d-Vectors and p-Vectors

Multiplying this by μ^{kj} and sum over k to get

$$\sum_i \underline{d}^{ik}\mu^{kj} = \sum_k(\underline{\rho}^r - \underline{\rho}^j)\left(\delta^{kj} - \frac{1}{m}m^k\right).$$

The left-hand side is $-\underline{b}^{ij}$. The right-hand side is $\underline{\rho}^r - \underline{\rho}^k - -\frac{1}{m}\underline{\rho}^r\sum_k m^k + \frac{1}{m}\sum_k m^i\underline{\rho}^k = -\underline{\rho}^k$. Finally

$$\underline{\rho}^k = \sum_i \underline{b}^{ik} . \tag{6}$$

At this point we can rewrite Eq. 1 as

$$\frac{d}{dt}(\mathbf{I}^i\cdot\underline{\omega}^i + \underline{h}^i) = \underline{L}^i + S^{ia}\underline{\mathcal{L}}^a + \underline{C}^{ia}S^{*as} \times\left[M^{sj}\sum_k \ddot{\underline{b}}^{kj} - \mu^{sj}\underline{F}^j\right]. \tag{7}$$

But note that $\underline{C}^{ia}S^{*as} = \underline{d}^{is}$ and that, by Eq. 4, $\underline{d}^{is}\mu^{sj} = -\underline{b}^{ij}$. Defining an equivalent external torque on the ith body as

$$\hat{\underline{L}}^i = \underline{L}^i + \underline{b}^{ij}\times\underline{F}^j . \tag{8}$$

Eq. 7 becomes

$$\frac{d}{dt}(\mathbf{I}^i\cdot\underline{\omega}^i + \underline{h}^i) = \hat{\underline{L}}^i + S^{ia}\underline{\mathcal{L}}^a + \sum_{j,k} m^j\underline{d}^{ij}\times\ddot{\underline{b}}^{kj} . \tag{9}$$

The further simplification of Eq. 9 is the subject of the next lecture.

13. Matrix Rotational Equations

We take up here with Eq. 9 of Lecture 12, and consider the last term

$$\sum_{j,k} m^j \underline{d}^{ij} \times \ddot{\underline{b}}^{kj} .$$

Consider first the case $i \neq k$. Suppose initially that either $v_k < v_i$ or that neither v_i nor v_k lie on the direct chain from the other to v_r. Then the terms in the sum are carried out only for $v_i \leq v_j$, because $\underline{d}^{ij} = 0$ in the remaining terms, so in all terms $\ddot{\underline{b}}^{kj} = \ddot{\underline{b}}^{ki}$. The sum over j therefore becomes

$$\sum_j m^j \underline{d}^{ij} \times \ddot{\underline{b}}^{kj} = \Big(\sum_j m^j \underline{d}^{ij}\Big) \times \ddot{\underline{b}}^{ki} = m\underline{b}^{ir} \times \ddot{\underline{b}}^{ki} .$$

But under the assumed conditions $\underline{b}^{ir} = \underline{b}^{ik}$ and we establish the result

$$\sum_j m^j \underline{d}^{ij} \times \ddot{\underline{b}}^{kj} = m\underline{b}^{ik} \times \ddot{\underline{b}}^{ki} . \tag{1}$$

Next, suppose that $v_i < v_k$. In this case we decompose the sum in two parts, the chain in which v_k is located and all other chains emanating from v_i except the direct chain from v_i to v_r . Let the body indices sets for these two cases be denoted Σ_1 and Σ_2 respectively, and write

$$\sum_j m^j \underline{d}^{ij} \times \ddot{\underline{b}}^{kj} = \sum_{j \in \Sigma_1} m^j \underline{d}^{ij} \times \ddot{\underline{b}}^{kj} + \sum_{j \in \Sigma_2} m^j \underline{d}^{ij} \times \ddot{\underline{b}}^{kj} .$$

In the first sum $\underline{d}^{ij} = \underline{d}^{ik}$ while in the second, $\ddot{\underline{b}}^{kj} = \ddot{\underline{b}}^{ki}$. Furthermore, because $\sum_j m^j \underline{b}^{kj} = 0$,

$$\sum_{j \in \Sigma_1} m^j \underline{b}^{kj} = -\sum_{j \notin \Sigma_1} m^j \underline{b}^{kj} = -\sum_{j \notin \Sigma_1} m^j \underline{b}^{ki} = -\Big(m - \sum_{j \notin \Sigma_1} m^j\Big)\underline{b}^{ki} .$$

It follows that

$$\begin{aligned}
\sum_j m^j \underline{d}^{ij} \times \ddot{\underline{b}}^{kj} &= \underline{d}^{ik} \times \Big(\sum_{j \in \Sigma_1} m^j - m\Big) \times \ddot{\underline{b}}^{ki} + \Big(\sum_{j \in \Sigma_2} m^j \underline{d}^{ij}\Big) \times \ddot{\underline{b}}^{ki} \\
&= \Big(\sum_j m^j \underline{d}^{ij} - m\underline{d}^{ik}\Big) \times \ddot{\underline{b}}^{ki} \\
&= m(\underline{b}^{ir} - \underline{d}^{ik}) \times \ddot{\underline{b}}^{ki} \\
&= m\underline{b}^{ki} \times \ddot{\underline{b}}^{ki} .
\end{aligned}$$

The result is the same as given by Eq. 1. Because all possible relationships between v_i and $v_k (i \neq k)$ now have been considered, that equation is established for all $k \neq i$.

If $k = i$, on the other hand,

$$\begin{aligned}
\sum_j m^j \underline{d}^{ij} \times \ddot{\underline{b}}^{ij} &= \sum_{v_i \leq v_j} m^j \underline{d}^{ij} \times \ddot{\underline{b}}^{ij} \\
&= \sum_{v_i \leq v_j} m^j (\underline{b}^{ir} - \underline{b}^{ij}) \times \ddot{\underline{b}}^{ij} \\
&= \underline{b}^{ir} \times \Big(-\sum_{v_i \nleq v_j} m^j \ddot{\underline{b}}^{ij}\Big) - \sum_{v_i \leq v_j} m^j \underline{b}^{ij} \times \ddot{\underline{b}}^{ij} .
\end{aligned}$$

But for $v_i \leq v_j$, $\underline{b}^{ir} = \underline{b}^{ij}$ and

$$\sum_j m^j \underline{d}^{ij} \times \ddot{\underline{b}}^{ij} = -\sum_j m^j \underline{b}^{ij} \times \ddot{\underline{b}}^{ij} . \tag{2}$$

Using the results given by Eqs. 1 and 2 in Eq. 9 of Lecture 12, the rotational equation for the ith body is

$$\frac{d}{dt}(\mathbf{I}^i \cdot \underline{\omega}^i + \underline{h}^i) = \hat{\underline{L}}^i + S^{ia}\underline{\mathcal{L}}^a - \sum_j m^j \underline{b}^{ij} \times \ddot{\underline{b}}^{ij} + \sum_{k \neq i} m\underline{b}^{ik} \times \ddot{\underline{b}}^{ki} . \tag{3}$$

Recall the definition of the augmented i^{th} body, and denote its inertia dyadic with respect to its barycenter as $\mathbf{K}^{ii}$. It is straightforward to show that

$$\mathbf{K}^{ii} = \mathbf{I}^i + \sum_j m^j (\underline{b}^{ij} \cdot \underline{b}^{ij} E - \underline{b}^{ij}\underline{b}^{ij}) \tag{4}$$

whence the first sum on the right hand side of Eq. 3 is

$$\begin{aligned} \sum_j m^j \underline{b}^{ij} \times \ddot{\underline{b}}^{ij} &= \frac{d}{dt}\left[\sum_j m^j \underline{b}^{ij} \times (\overset{\circ}{\underline{b}}{}^{ij} + \underline{\omega}^i \times \underline{b}^{ij})\right] \\ &= \frac{d}{dt}\left[\sum_j m^j \underline{b}^{ij} \times \overset{\circ}{\underline{b}}{}^{ij} + (\mathbf{K}^{ii} - \mathbf{I}^i)\cdot \underline{\omega}^i\right] . \end{aligned} \tag{5}$$

The relative derivative $\overset{\circ}{\underline{b}}{}^{ij}$ is with respect to the X^i-frame. If the body is rigid or a rigid gyrostat, that relative derivative vanishes.

Introducing

$$\underline{H}^i = \underline{h}^i + \sum_j m^j \underline{b}^{ij} \times \overset{\circ}{\underline{b}}{}^{ij} \tag{6}$$

and using Eqs. 5 and 6 in Eq. 3, the latter becomes

(7) $$\frac{d}{dt}(K^{ii}\cdot\underline{\omega}^{i} + \underline{H}^{i}) = \hat{\underline{L}}^{i} + S^{ia}\underline{\mathfrak{L}}^{a} + \sum_{k\neq i} m\underline{b}^{ik} \times \ddot{\underline{b}}^{ki} .$$

It is important to recognize that this form of the rotational e quation holds for non-rigid as well as rigid systems.

This can be written as a matrix equation by inspection. Resolving all i^{th} body equations in the frame X^{i}, we get

(8) $$\frac{d}{dt}(K^{ii}\omega^{i} + H^{i}) + \tilde{\omega}^{i}(K^{ii}\omega^{i} + H^{i}) = \hat{L}^{i} + S^{ia}\underline{\mathfrak{L}}^{a}\cdot\underline{X}^{i}$$
$$+ \sum_{k\neq i} m\tilde{b}^{ik}A^{ik}\left[\ddot{b}^{ki} + 2\tilde{\omega}^{k}\dot{\underline{b}}^{ki} + (\dot{\tilde{\omega}}^{k} + \tilde{\omega}^{k}\tilde{\omega}^{k})b^{ki}\right] .$$

We now subject Eq. 8 to Hooker's reduction, some what modified in this presentation. It is based on the fact that in most problems it is more accurate, numerically speaking, and more in accordance with the needs of application if one refers the motion of all bodies to that of the reference body. Thus the absolute angular velocities $\underline{\omega}^{i}$ of the several bodies are replac ed by the angular velocity $\underline{\omega}$ of the reference body plus a suita ble relative angular velocity. We define a relative angular velocity $\underline{\Omega}^{a}$ in each arc as

(9) $$\underline{\Omega}^{a} = S^{ia}\underline{\omega}^{i}$$

(twice-repeated superscripts are summed). Multiply this by S^{*aj} and sum over a, using the fact that

$$S^{ia}S^{*aj} = \delta^{ij} - \delta^{ir} \tag{10}$$

in terms of Kronecker deltas. Then it follows immediately that

$$\underline{\omega}^{j} = \underline{\omega} + S^{*aj}\underline{\Omega}^{a}. \tag{11}$$

To prove Eq. 10, suppose $v_i \nleq v_j$. Then $S^{ia} \neq 0$ only if a is the arc toward the reference body, but $S^{*aj} \neq 0$ then, because a is outboard of body j. Hence $S^{ia}S^{*aj} = 0$ in this case. If $v_i = v_j \neq v_r$ the sum is over the arcs of the ith body; but of these only the one has a non-zero S^{*}-value, namely the one toward the reference body. For this arc S^{ia} and S^{*ai} have the same sign, so $S^{ia}S^{*aj} = 1$. If $v_i < v_j$, say v_i on the chain to v_i at a_1, the non-zero terms of the sum are those for the arcs a_1 and a_r (toward the reference body). For a_1, S^{ia_1} and S^{*a_1j} have opposite signs, whereas for a_r the signs are the same. It follows that $S^{ia}S^{*aj} = 0$. If $v_j = v_r$, $S^{*aj} = 0$ and the sum vanishes. These properties are embodied in Eq. 10.

In matrix form we write Eq. 11 as

$$\omega^{j} = A^{jr}(\omega + S^{*aj}\Omega^{a}) \tag{12}$$

where $\underline{\Omega}^{a}$ are resolved in the reference frame. Also

$$\dot{\omega}^{j} = A^{jr}(\dot{\omega} + S^{*aj}\dot{\Omega}^{a} + \tilde{\omega}S^{*aj}\Omega^{a}). \tag{13}$$

Let Eq. 8 be differentiated and Eq. 13 used to replace $\dot{\omega}^i$ in the result. We get

$$(14)\qquad K^{ii}A^{ir}(\dot{\omega} + S^{*ai}\dot{\Omega}^a) + \sum_{k\neq i} m\tilde{\underline{b}}^{ik}A^{ik}\tilde{b}^{ki}A^{kr}(\dot{\omega} + S^{*ak}\dot{\Omega}^a) = \mathfrak{C}_0^i + S^{ia}\underline{\mathfrak{L}}^a\cdot\underline{X}^i$$

where, for brevity, we have let $\mathfrak{C}_0^i$ stand for

$$(15)\qquad \mathfrak{C}_0^i = -\dot{K}^{ii}\omega^i - \dot{H}^i - K^{ii}A^{ir}\tilde{\omega}S^{*ai}\Omega^a - \tilde{\omega}^i(K^{ii}\omega^i + \dot{H}^i) + \hat{L}^i + \sum_{k\neq i} m\tilde{b}^{ik}A^{ik}\left[\ddot{b}^{ki} + 2\tilde{\omega}^k\dot{b}^{ki} + \tilde{\omega}^k\tilde{\omega}^k b^{ki} - \tilde{b}^{ki}A^{kr}\tilde{\omega}S^{*ak}\Omega^a\right].$$

Next, define

$$(16)\qquad \Phi^{ik} = \begin{cases} K^{ii} & (k = i) \\ m\tilde{b}^{ik}A^{ik}\tilde{b}^{ki} & (k \neq i). \end{cases}$$

Equation 14 then becomes

$$(17)\qquad \sum_k \Phi^{ik}A^{kr}\dot{\omega} + \sum_a\sum_k S^{*ak}\Phi^{ik}A^{kr}\dot{\Omega}^a = \mathfrak{C}_0^i + S^{ia}\underline{\mathfrak{L}}^a\cdot\underline{X}^i.$$

Now consider the rotational degrees of freedom in the ath arc. Let vector array $\underline{u}^a$ have one, two or three unit-vector elements which comprise a minimal set spanning the modes of possible angular motion in the ath arc. We can define a matrix Θ^a of the same order as the vector array, representing rotation angles about the elements of $\underline{u}^a$. Then

$$\underline{\Omega}^a = \underline{u}^{a^T} \dot{\theta}^a \qquad \text{(not summed)}. \tag{18}$$

If $\underline{u}^a$ be resolved in the reference body as matrix U^a , then

$$\dot{\Omega}^a = U^{a^T} \ddot{\theta}^a + \dot{U}^{a^T} \dot{\theta}^a \text{ (not summed)}. \tag{19}$$

Further, define

$$\mathcal{L}^b = \underline{u}^b \cdot \underline{\mathcal{L}}^b \tag{20}$$

as a matrix of torque components on the unit vectors spanning the free hinge modes, and

$$\mathfrak{C}^i = \mathfrak{C}_0^i - \sum_{a,k} S^{*ak} \Phi^{ik} \dot{U}^{a^T} \dot{\theta}^a . \tag{21}$$

Using Eqs. 19 and 21 in Eq. 17, and temporarily restoring its vectorial form, the latter becomes

$$\underline{X}^{i^T} \left[\sum_k \Phi^{ik} A^{kr} \dot{\omega} + \sum_{a,k} S^{*ak} \Phi^{ik} A^{kr} U^{a^T} \ddot{\theta}^a \right] = \underline{X}^{i^T} \mathfrak{C}^i + S^{ia} \underline{\mathcal{L}}^a . \tag{22}$$

We now take two steps. First we simply sum Eq. 22 over i, in which process all terms in $\underline{\mathcal{L}}^a$ disappear, and restore the matrix form by resolving in the reference frame. The result is

$$\sum_{i,k} (A^{ri} \Phi^{ik} A^{kr}) \dot{\omega} + \sum_a \left(\sum_{i,k} S^{*ak} A^{ri} \Phi^{ik} A^{kr} U^{a^T} \right) \ddot{\theta}^a = \sum_i A^{ri} \mathfrak{C}^i . \tag{23}$$

Second, we convert $\underline{X}^i$ to $\underline{X}^r$ in Eq. 22, multiply by $S^{*bi} \underline{u}^b$ and sum over i , to get (not summed on b)

$$\left(\sum_{i,k} S^{*bi} U^b A^{ri} \Phi^{ik} A^{kr}\right)\dot{\omega} + \sum_a \left(\sum_{i,k} S^{*bi} U^b A^{ri} \Phi^{ik} A^{kr} U^{a^T} S^{*ak}\right)\ddot{\theta}^a =$$

(24)
$$= \sum_i S^{*bi} U^b A^{ri} \mathfrak{C}^i + \sum_i S^{*bi} S^{ia} \underline{u}^b \cdot \underline{\mathfrak{L}}^a .$$

Put

(25a)
$$J^{oo} = \sum_{i,k} (A^{ri} \Phi^{ik} A^{kr})$$

(25b)
$$J^{oa} = \sum_{i,k} S^{*ak} (A^{ri} \Phi^{ik} A^{kr}) U^{a^T}$$

(25c)
$$J^{bo} = \sum_{i,k} S^{*bi} U^b (A^{ri} \Phi^{ik} A^{kr})$$

(25d)
$$J^{ba} = \sum_{i,k} S^{*bi} U^b (A^{ri} \Phi^{ik} A^{kr}) U^{a^T} S^{*ab} .$$

Also

(26a, b)
$$\mathfrak{D} = \begin{bmatrix} \sum_i A^{ri} \mathfrak{C}^i \\ \hdashline \left[\sum_i S^{*bi} U^b A^{ri} \mathfrak{C}^i\right] \end{bmatrix}, \quad \underline{\mathfrak{L}} = \begin{bmatrix} 0 \\ \mathfrak{L}^1 \\ \vdots \\ \mathfrak{L}^{n-1} \end{bmatrix}$$

where the lower partitioning of $\mathfrak{D}$ is a column matrix whose row index is b. Then Eqs. 23 and 24 together take the very neat form

(27)
$$\begin{bmatrix} J^{00} & J^{01} & J^{02} \dots \\ J^{10} & J^{11} & J^{12} \dots \\ J^{20} & J^{21} & J^{22} \dots \\ & \dots & \end{bmatrix} \begin{bmatrix} \dot{\omega} \\ \ddot{\theta}^1 \\ \vdots \\ \ddot{\theta}^{n-1} \end{bmatrix} = \mathfrak{D} + \mathfrak{L} .$$

One of the nicest features of this formalism is that the J-matrix is symmetric. This is easily proved from Eqs. 25 using the fact that $\Phi^{ik^T} = \Phi^{ki}$.

If any of rows of $\mathcal{L}$ represent rotational modes in the arcs where a purely kinematical constraints is imposed, that row is stricken from Eq. 27 and the terms from the corresponding column of the J-matrix are transferred to the right as known driving terms.

14. Applications

The dynamical formalism for systems of interconnected rigid bodies developed in the previous lectures can be applied to mechanical problems of very different nature. When it was first developed in 1965 its authors had in mind an application to spacecraft. Some details of the mathematics in this case and a typical result will be shown after this introduction. Another filed for applications is found in the study of mechanisms. One aspect will be described as well in this lecture. There are also systems which can be called living mechanisms. The human body, as well as any other animal body, is a system with tree-structure made up of interconnected bodies which in many cases can be considered approximately rigid. It is interesting to note that in 1905 a German mathematician called Fischer investigated the dynamics of the human body [1]. He discovered the augmented-body concept and some other important relationships. Due to an unfortunate choice of variables, however, he did not arrive at a convenient form for the equations of motion. He considered the problem: Given the motion of the human body as a function of time, i.e. the angular positions, velocities and accelerations of its parts, what are the muscle forces as functions of time necessary to produce this motion? He actually kinematographically recorded the motion of a walking person and numerically determined muscle forces acting in the legs. Another possible application to the human body is the determination of its motion relative to a car immediately following a car collision.

In the following, two applications will be demonstrated in some detail. In the first one a multi-body satellite with tree-structure and with spherical hinges only is investigated which is moving in a circular orbit about Earth. The gravitational attraction being a function of the distance from the center of Earth produces torques which tend to rotate the satellite unless it is in an equilibrium attitude in which the gravitational torques, the torques due to centrifugal forces and the torques produced by the hinge reaction forces cancel each other on each individual body of the system. To be determined are these particular equilibrium attitudes. To do this an expression has to be found for the term $\underline{L} + \underline{P}\,\underline{F}$ in the equations of motion. Fig. 1 shows a typical satellite in an orbit of radius r_0. The gravitational force on body i is

$$\vec{F}^{i} = -\varkappa \int_{m^{i}} \frac{(\vec{r}_{o} + \vec{R}^{i} + \vec{\varrho})}{|\vec{r}_{o} + \vec{R}^{i} + \vec{\varrho}|^{3}} dm$$

where $\vec{\varrho}$ is the radius vector of a particle dm with respect to the body i center of mass and $\varkappa$ is the product of Earth mass and gravitational constant. Neglecting terms of higher than first order in R^2/r_o and ϱ/r_o we can write

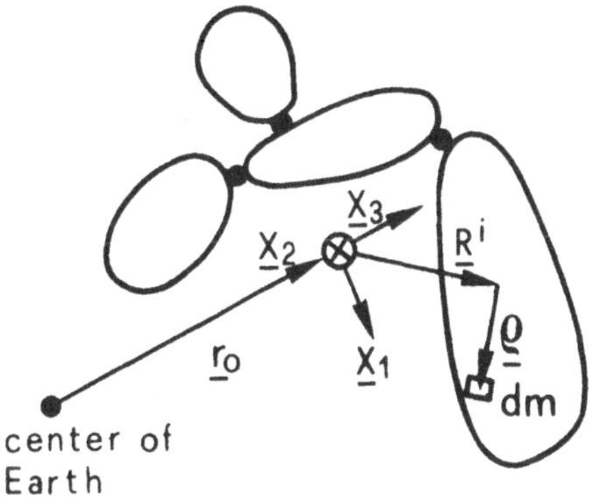

Fig. 1

$$\begin{aligned}
\vec{F}^{i} &\approx -\frac{\varkappa}{r_o^3} \int_{m^{i}} \frac{\vec{r}_o + \vec{R}^2 + \vec{\varrho}}{1 + 3\vec{r}_o\cdot(\vec{R}^{i} + \vec{\varrho})/r_o^2} dm \\
&\approx -\frac{\varkappa}{r_o^3} \int_{m^{i}} (\vec{r}_o + \vec{R}^{i} + \vec{\varrho})[1 - 3\vec{r}_o\cdot(\vec{R}^{i} + \vec{\varrho})/r_o^2] dm \qquad (1) \\
&\approx -\frac{\varkappa m^{i}}{r_o^3} [\vec{r}_o + \vec{R}^{i} - 3(\vec{r}_o\cdot\vec{R}^{i})\vec{r}_o] \, .
\end{aligned}$$

Similarly, the gravitational torque is

$$\vec{L}^{i} = \varkappa \int_{m^{i}} \frac{\vec{\varrho} \times (\vec{r}_o + \vec{R}^{i} + \vec{\varrho})}{|\vec{r}_o + \vec{R}^{i} + \vec{\varrho}|^{3}} dm$$

$$\approx \frac{3\varkappa}{r_o^5}\vec{r}_o \times \int_{m^i} \vec{\varrho}(\vec{r}_o \cdot \vec{\varrho})dm \; .$$

Using the identity

$$\vec{\varrho}(\vec{r}_o \cdot \vec{\varrho}) = -\vec{\varrho} \times (\vec{\varrho} \times \vec{r}_o) + \vec{\varrho}^{\,2}\vec{r}_o$$

this becomes

$$\vec{L}^i \approx \frac{3\varkappa}{r_o^5}\vec{r}_o \times \int_{m^i} [-\vec{\varrho} \times (\vec{\varrho} \times \vec{r}_o)]dm \; .$$

Decomposition into body-fixed components is most easily carried out by introducing the so-called orbital coordinate system $\vec{x}_{1,2,3}$ shown in Fig. 1. Its origin is always coincident with the satellite center of mass. Its unit vectors $\vec{x}_1, \vec{x}_2, \vec{x}_3$ are pointing in the direction of the orbit tangent (forward), of the orbital angular velocity $\vec{\omega}_e$ and of the local vertical (upward), respectively. This coordinate system is rotating in inertial space with $\vec{\omega}_o$. In the equations of motion there appeared in various terms the coordinate transformation matrices $\underline{A}^{ij}$ relating body-fixed coordinate systems of different bodies i and j. We now write $\underline{A}^{ij} = \underline{A}^i \underline{A}^{j^T}$ and assume $\underline{A}^i$ as coordinate transformation matrix between the orbital frame of reference and the body i frame of reference, i.e. $\underline{\vec{e}}^i = \underline{A}^i \underline{\vec{x}}$ with $\underline{\vec{x}} = [\vec{x}_1 \, \vec{x}_2 \, \vec{x}_3]^T$. Denoting by $\underline{A}_2^i$ and $\underline{A}_3^i$ the second and third columns of $\underline{A}^i$ the components of $\vec{L}^i$ in $\underline{\vec{e}}^i$ are ob tained as

$$\underline{L}^{i} = \frac{3\varkappa}{r_o^3}\tilde{\underline{A}}_3^{i}\left(-\int_{m^i}\tilde{\underline{\varrho}}\tilde{\underline{\varrho}}\,dm\right)\underline{A}_3^{i}$$

where $\tilde{\underline{\varrho}}$ is made up of body-fixed components of $\vec{\varrho}$. The term in brackets is the inertia matrix $\underline{I}^{i}$ as is easily seen by multiplying out the terms $\bar{\varrho}\,\bar{\varrho}$. Thus

$$\underline{L}^{i} = \frac{3\varkappa}{r_o^3}\tilde{\underline{A}}_3^{i}\underline{I}^{i}\underline{A}_3^{i} \; .$$

The term $\underline{L}$ in the equations of motion, finally, becomes

$$\underline{L} = \frac{3\varkappa}{r_o^3}\tilde{\underline{A}}_3\underline{I}\,\underline{A}_3$$

where $\underline{A}_3$ is a column matrix containing all $\underline{A}_3^{i}$ and $\tilde{\underline{A}}_3$ is a $(3n\times 3n)$-matrix with submatrices $\delta_{ij}\tilde{\underline{A}}_3^{i}(i,j = 1,\ldots,n)$. In order to construct the product $\underline{P}\underline{F}$ first the column matrix $[\vec{F}^{i}]$ is developed. Eq. (1) yields

$$\vec{\underline{e}}^{T}\underline{F} = [\vec{F}^{i}] = -\frac{\varkappa}{r_o^3}\left\{\vec{r}_o\underline{m}\underline{1}_n + \underline{m}[\vec{R}^{i}] - \frac{3}{r_o^2}\underline{m}\vec{r}_o\left([\vec{R}^{i}]\circ\vec{r}_o\right)\right\}$$

or with theorem of Lecture N° 12

$$[\vec{F}^{i}] = -\frac{\varkappa}{r_o^3}\left\{\vec{r}_o\underline{m}\underline{1}_n - \underline{m}\underline{\mu}^{T}\underline{T}^{T}\underline{C}^{T}\left[\vec{\underline{e}}\,\underline{1}_n - \frac{3}{r_o^2}\vec{r}_o(\vec{\underline{e}}\,\underline{1}_n\circ\vec{r}_o)\right]\right\}$$

and with the identities

$$\vec{\underline{e}}\,\underline{1}_n\circ\vec{r}_o = r_o\underline{A}_3\,, \quad \vec{r}_o(\vec{\underline{e}}\,\underline{1}_n\circ\vec{r}_o) = r_o^2\left[\underline{A}_3^i\underline{A}_3^{i^T}\,\vec{\underline{e}}^{\,i}\right]$$

finally

$$(2) \qquad \left[\vec{F}^i\right] = \frac{\varkappa}{r_o^3}\left\{-\vec{r}_o\underline{m}\,\underline{1}_n + \underline{m}\underline{\mu}^T\,\underline{I}^T\underline{C}^T\left[(\underline{E}_3 - 3\underline{A}_3^i\underline{A}_3^{i^T})\vec{\underline{e}}^{\,i}\right]\right\}.$$

Instead of premultiplying this scalarly with $\vec{\underline{e}}$ in order to obtain $\underline{F}$ and then to carry out the multiplication $\underline{P}\underline{F}$ it is more convenient to go back to the equation

$$\underline{P}\underline{F} = -\vec{\underline{e}}\circ\vec{\underline{e}}^T\otimes\underline{C}\,\underline{T}\,\underline{\mu}\,\underline{e}^T\underline{F}\,.$$

Substituting (2) one obtains

$$\underline{P}\underline{F} = -\frac{\varkappa}{r_o^3}\,\vec{\underline{e}}\circ\vec{\underline{e}}\otimes\underline{J}\left[(\underline{E}_3 - 3\underline{A}_3^i\underline{A}_3^{i^T})\vec{\underline{e}}^{\,i}\right]$$

where $\underline{J}$ is the matrix of the same name defined in lecture N° 13. It becomes apparent here that $\underline{L} + \underline{P}\underline{F}$ can be expressed in terms of augmented-body parameters. Rather straight-forward manipulations (found in detail in [2], [3]) lead to the result

$$\underline{L} + \underline{P}\underline{F} = \frac{\varkappa}{r_o^3}\left\{3\tilde{\underline{A}}_3\underline{K}\underline{A}_3 + M\left[\sum_{\substack{j=1\\ j\neq i}}^{n}\tilde{\underline{B}}_{ij}(\underline{A}^{ij} - 3\underline{A}_3^i\underline{A}_3^{j^T})\underline{B}_{ji}\right]\right\}.$$

This has to be substituted together with $\left[\vec{Y}_a\right] = 0$ into the equations of motion. The desired equilibrium positions are character ized by the fact that no motion relative to the orbital coordinate system occurs. Thus $\dot{\underline{\omega}}^{i} = 0$ and $\underline{\omega}^{i} = \underline{\omega}_c^{i}\underline{A}_2^{i} (i = 1,\dots,n)$. Furthermore, in circular orbits $3\varkappa/r_o^3 = \omega_o^2$. The equations of motion then reduce to equilibrium conditions of the form

$$\tilde{\underline{A}}_2\underline{K}\underline{A}_2 - M\left[\sum_{\substack{j=1\\ j\neq i}}^{n}\tilde{\underline{B}}_{ij}\underline{A}^{ij}\tilde{\underline{A}}_2^{j}\tilde{\underline{A}}_2^{j}\underline{B}_{ji}\right] =$$

$$= 3\tilde{\underline{A}}_3\underline{K}\underline{A}_3 + M\left[\sum_{\substack{j=1\\ j\neq i}}^{n}\tilde{\underline{B}}_{ij}(\underline{A}^{ij} - 3\underline{A}_3^{i}\underline{A}_3^{j\,T})\underline{B}_{ji}\right].$$

Recognizing the identity

$$\underline{A}^{ij}\tilde{\underline{A}}_2^{j}\tilde{\underline{A}}_2^{j} = \underline{A}^{i}\underline{A}^{j\,T}(\underline{A}_2^{j}\underline{A}_2^{j\,T} - \underline{E}_3) = \underline{A}_2^{i}\underline{A}_2^{j\,T} - \underline{A}^{ij}$$

this can be given the form

$$\tilde{\underline{A}}_2\underline{Q}\underline{A}_2 = 3\underline{A}_3\underline{Q}\underline{A}_3 \qquad (3)$$

where $\underline{Q}$ is a constant symmetric $(3n \times 3n)$-matrix with (3×3)-submatrices

$$\underline{Q}_{ij} = \begin{cases} \underline{K}^{i} & i = j \\ M\underline{B}_{ij}\underline{B}_{ji}^{T} & i \neq j\,. \end{cases}$$

Eq. (3) is a set of $3n$ algebraic equations for elements of the transformation matrices $\underline{A}^{1},\dots,\underline{A}^{n}$ which determine the equilibrium positions of the system. They have to be supplemented by the orthonormality conditions

$$\underline{A}_2^{i\,T}\underline{A}_2^{i} = 1\,, \quad \underline{A}_3^{i\,T}\underline{A}_3^{i} = 1\,, \quad \underline{A}_2^{i\,T}\underline{A}_3^{i} = 0\,, \quad i = 1,\dots,n\,. \qquad (4)$$

The Eqs. (3) and (4) are coupled second-order equations. There is no mathematical theory which allows to make any statement about the number of solutions, about the number of real solutions or about the solutions themselves. Therefore, analytical results can be found only for some classes of satellite systems for which the matrix $\underline{Q}$ has a particularly simple form, i.e. where it has only few non-zero elements. Only one special example will be given here. For more results the reader is referred to [2] . Let us consider a satellite consisting of two homogeneous, brick-shaped bodies which are coupled in the midpoint of one of their edges (Fig. 2). Furthermore, we restrict our interest to equilibrium positions where the two edges carrying the hinge are perpendicular to the orbit plane. In this case the equilibrium positions are defined by two angles φ_1 and φ_2 specifying the direction of the body-fixed axes $\vec{e}_1^{\,1}$ and $\vec{e}_1^{\,2}$ with respect to $\vec{x}_1$. The matrices $\underline{A}^1$ and $\underline{A}^2$ have the form

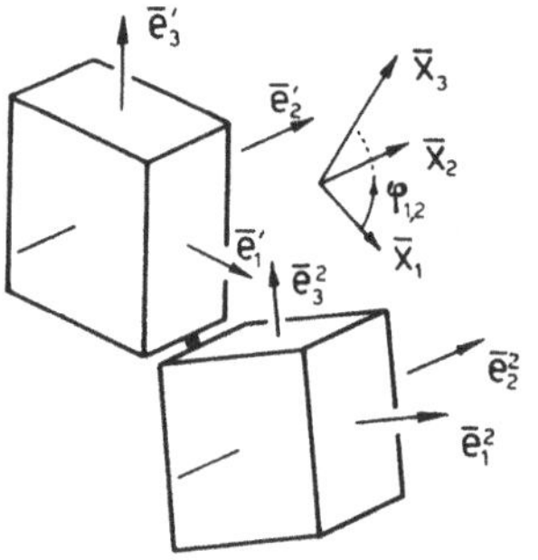

Fig. 2

$$\underline{A}^i = \begin{bmatrix} \cos\varphi_i & 0 & \sin\varphi_i \\ 0 & 1 & 0 \\ -\sin\varphi_i & 0 & \cos\varphi_i \end{bmatrix}, \quad i = 1,2$$

yielding

$$\underline{\tilde{A}}_2^i = \begin{bmatrix} 0 & 0 & 1 \\ 0 & 0 & 0 \\ -1 & 0 & 0 \end{bmatrix}, \quad \underline{\tilde{A}}_3^i = \begin{bmatrix} 0 & -\cos\varphi_i & 0 \\ \cos\varphi_i & 0 & -\sin\varphi_i \\ 0 & \sin\varphi_i & 0 \end{bmatrix}, \quad i = 1,2 .$$

The coordinate axes $\vec{e}_1^{\,1}$ and $\vec{e}_2^{\,2}$ being principal axes of the individual as well as of the augmented bodies we have

$$\underline{K}^i = \begin{bmatrix} K_{11} & 0 & -K_{13} \\ 0 & K_{22} & 0 \\ -K_{13} & 0 & K_{33} \end{bmatrix}^i, \quad i = 1,2 .$$

Furthermore, with $\underline{B}_{ij} = \begin{bmatrix} B_{ij1} & 0 & B_{ij3} \end{bmatrix}^\tau$ the submatrices of the matrix Q for $i \neq j$ are

$$\underline{Q}_{ij} = \begin{bmatrix} Q_{ij11} & 0 & Q_{ij13} \\ 0 & 0 & 0 \\ Q_{ij31} & 0 & Q_{ij33} \end{bmatrix} .$$

Multiplying out Eq. (3) one gets two equations for the unknowns φ_1 and φ_2 which read

$$(K_{33}^i - K_{11}^i)\sin\varphi_i\cos\varphi_i - K_{13}^i(\sin^2\varphi_i - \cos^2\varphi_i) -$$

$$-\cos\varphi_i(Q_{ij11}\sin\varphi_j + Q_{ij13}\cos\varphi_j) + \sin\varphi_i(Q_{ij31}\sin\varphi_j + Q_{ij33}\cos\varphi_j) = 0$$

$$i,j = 1,2\,, \quad j \neq i\,. \tag{5}$$

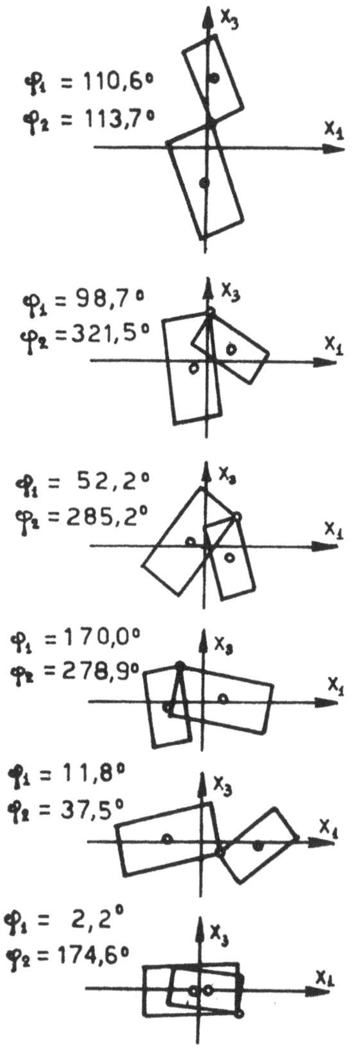

Fig. 3

The orthonormality conditions are automatically satisfied by the special form of $\underline{A}^1$ and $\underline{A}^2$. Eq. (5) was numerically solved for the case of two bodies with side lengths of 8 by 4 by 1 units of length (body 1) and 6 by 3 by 1 units (body 2) the lengths being given in the sequence $\vec{e}_{1,2,3}$ (Fig. 2). With these parameters (5) has twelve real solutions six of which are shown in Fig. 3. The remaining ones are obtained by a 180° - rotation of the entire system about the axis $\vec{x}_2$. A stability analysis reveals that only the equilibrium position with $\varphi_1 = 110{,}6°$ and $\varphi_2 = 113{,}7°$ is stable. For practical applications only stable equilibrium positions are of interest. It is, therefore, desirable to have an algorithm furnishing only these positions. Two different such algorithms have been developed ([4], [7]).

The second application of the dynamical formalism is dealing with impact problems in mechanisms. An impact $\hat{\vec{F}}$ acting in some point of a multi-body system is understood as a Dirac pulse $\hat{\vec{F}} = \lim_{\Delta t \to 0} \int_0^{\Delta t} \vec{F}(t)dt$ with an infinitely large force $\vec{F}(t)$ acting over an infinitesimally small time interval Δt. It causes a sudden finite change of velocities and angular velocities whereas the position of the system remains constant during the time interval Δt. If a tree-structured system with frictionless spherical hinges is subject to simultaneously applied impacts $\hat{\vec{F}}^i$ $(i = 1, 2, \ldots, n)$ where the index i is denoting the body the following equations of motion can be formulated

$$\begin{aligned} \underline{m}[\Delta\dot{\vec{R}}^i] &= \underline{\mu}[\hat{\vec{F}}^i] + \underline{S}[\hat{\vec{X}}_a] \\ [\overrightarrow{\Delta H}^i] &= [\hat{\vec{L}}^i] + \underline{\vec{e}}^T \underline{C} \otimes [\hat{\vec{X}}_a] \;. \end{aligned} \tag{6}$$

They constitute integrals over $\Delta t \to 0$ of the equations of motion

$$\begin{aligned} \underline{m}[\ddot{\vec{R}}^i] &= \underline{\mu}[\vec{F}^i] + \underline{S}[\vec{X}_a] \\ [\vec{H}^i] &= [\vec{L}^i] + \underline{\vec{e}}^T \underline{C} \otimes [\vec{X}_a] \end{aligned} \tag{7}$$

which were the starting point for the development of the formalism for multy-body systems (see lecture N° 11). $\Delta\dot{\vec{R}}^i$ and $\overrightarrow{\Delta H}^i$ are the finite increments of $\dot{\vec{R}}^i$ and $\vec{H}^i$ during impact, $\hat{\vec{L}}^i$ is the torque caused by the pulse $\hat{\vec{F}}^i$ and $\hat{\vec{X}}_a$ is a hinge reaction pulse. From (6) follows in a way almost identical with the one leading from (7) to the equations of motion derived earlier

$$(\underline{K} + \underline{\Psi})\underline{\Delta\omega} = \underline{\hat{L}} + \underline{P}\underline{\hat{F}} \tag{8}$$

where $\underline{K}$, $\underline{\Psi}$ and $\underline{P}$ are the matrices defined in lecture N° 13. The column matrices $\underline{\hat{F}}$ and $\underline{\hat{L}}$ are containing components (measured in body-fixed coordinate systems) of pulses and torques caused by the pulses respectively. Let $\vec{\varrho}^{\,i}$ be the vector from the body i center of mass to the point of application of $\hat{\vec{F}}^{i}$ and let $\underline{\varrho}^{i}$ be the matrix of body-fixed components of $\vec{\varrho}^{\,i}$. Then the torque pulse $\hat{\vec{L}}^{i}$ is identical with $\vec{\varrho}^{\,i}\times\hat{\vec{F}}^{i}$ and its component presentation is $\underline{\tilde{\varrho}}^{i}\underline{\hat{F}}^{i}$. With this the right hand side of (8) can be written in the form

$$\underline{\hat{L}} + \underline{P}\underline{\hat{F}} = \underline{V}^{*}\underline{\hat{F}}$$

where $\underline{V}^{*}$ is a $(3n \times 3n)$-matrix with (3×3)-submatrices

$$\underline{V}^{*}_{ij} = \underline{\tilde{B}}_{ij}\underline{A}^{ij} + \delta_{ij}\underline{\tilde{\varrho}}^{i} = (\underline{\tilde{B}}_{ij}\underline{A}^{i} + \delta_{ij}\underline{\tilde{\varrho}}^{i}\underline{A}^{j})\underline{A}^{j^T} , \qquad i,j = 1,\dots,n .$$

Finally, combining the matrices $\underline{A}^{j^T}$ in this last expression with the corresponding factors $\underline{\hat{F}}^{j}$ one arrives at the equation

$$(\underline{K} + \underline{\Psi})\underline{\Delta\omega} = \underline{V}\,\underline{\hat{F}}_{abs} \tag{9}$$

where now $\underline{\hat{F}}_{abs}$ is containing the components of all n pulses in the common reference system $\underline{\vec{x}}$ whereas the matrix $\underline{V}$ has submatrices

$$\underline{V}_{ij} = \underline{\tilde{B}}_{ij}\underline{A}^i + \delta_{ij}\underline{\tilde{\varrho}}^i\underline{A}^j \quad , \qquad i,j = 1,\ldots,n . \tag{10}$$

With (9) and (10) the dynamics section of the impact problem has been completed. Next, the simpler kinematics section is developed. Let the symbols $\vec{v}_0$ and $\vec{v}^i$ denote the translational velocities of the system center of mass and of the point of application of $\hat{\vec{F}}^i$ on body i , respectively. Then the equations hold

$$\Delta\vec{V}_0 = \frac{1}{M}\sum_{j=1}^{n}\hat{\vec{F}}^j \tag{11}$$

$$\Delta\vec{V}^i = \Delta\vec{V}_0 + \Delta\dot{\vec{R}}^i - \vec{\varrho}^i \times \Delta\vec{\omega}^i \quad , \qquad i = 1,\ldots,n \tag{12}$$

where capital delta, as before, denotes the change of the respective quantity during impact. All n equations (12) are combined in the matrix equation

$$[\Delta\vec{V}^i] = \Delta\vec{V}_0\underline{1}_n + [\Delta\dot{\vec{R}}^i] - [\vec{\varrho}^i \times \Delta\vec{\omega}^i] . \tag{13}$$

The first term is with (11)

$$\Delta\vec{V}_0\underline{1}_n = \frac{1}{M}\underline{1}_n\underline{1}_n^T[\hat{\vec{F}}^i] .$$

The second term is with one of the basic theorems derived in Lecture N° 12

$$[\Delta\dot{\vec{R}}^i] = -[\sum_{j=1}^{n}\vec{B}_{ji} \times \Delta\vec{\omega}^j] .$$

Consequently, (13) becomes

$$(14) \qquad \left[\vec{\Delta V}^i\right] = \frac{1}{M}\underline{1}_n\underline{1}_n^T\left[\hat{\vec{F}}^i\right] - \left[\sum_{j=1}^n (\vec{B}_{ji} + \delta_{ij}\vec{\varrho}^i)\times\vec{\Delta\omega}^j\right].$$

The vectors $\vec{\Delta v}^i$ are now decomposed in the reference coordinate system $\underline{\vec{x}}$. Then, the left side of (14) yields a column matrix $\underline{\Delta v}_{abs}$ of $3n$ elements. The term involving $\left[\hat{\vec{F}}^i\right]$ contributes the expression $\frac{1}{M}\underline{D}\hat{\underline{F}}_{abs}$ with a $(3n \times 3n)$-matrix $\underline{D}$ whose (3×3)-submatrices are all unit matrices. The last expression in 14, finally, contributes the term

$$-\left[\sum_{j=1}^n \underline{A}^{j^T}(\underline{\tilde{B}}_{ji} + \delta_{ij}\underline{\tilde{\varrho}}^i)\underline{\Delta\omega}^j\right] =$$

$$= \left[\sum_{j=1}^n \{(\underline{\tilde{B}}_{ji} + \delta_{ij}\underline{\tilde{\varrho}}^i)\underline{A}^j\}^T\underline{\Delta\omega}^j\right].$$

The matrix $\underline{A}^j$ occurs because $\underline{B}_{ji}$, $\underline{\varrho}^i$ and $\underline{\Delta\omega}^j$ contain the components of the respective vectors in the system $\underline{\vec{e}}^j$. Eq. (14) now has the form

$$\underline{\Delta V}_{abs} = \frac{1}{M}\underline{D}\hat{\underline{F}}_{abs} + \left[\sum_{j=1}^n \{(\underline{\tilde{B}}_{ji} + \delta_{ij}\underline{\tilde{\varrho}}^i)\underline{A}^j\}^T\underline{\Delta\omega}^j\right].$$

Because of (10) this is identical with

$$\underline{\Delta V}_{abs} = \frac{1}{M}\underline{D}\hat{\underline{F}}_{abs} + \underline{V}^T\underline{\Delta\omega}.$$

Combined with (9) this yields the final result

$$\underline{\Delta V}_{abs} = \left\{ \frac{1}{M}\underline{D} + \underline{V}^T(\underline{K} + \underline{\Psi})^{-1}\underline{V} \right\} \hat{\underline{F}}_{abs} = \underline{G}\hat{\underline{F}}_{abs} \; . \qquad (15)$$

The fact that $\underline{G}$ is a symmetric matrix establishes the validity of the law of Maxwell and Betti for this dynamics problem and a mathematical analogy between impact dynamics of mechanisms and statics in linear elasticity. For more details and further generalizations the reader is referred to [5], [6] . The practical applicability of (15) will now be demonstrated by the simple example shown in Fig. 4. A system of eight bodies suspended in one point (all hinges including the support are spherical hinges) and initially at rest is hit by a point mass m with velocity v. To be determined are the augular velocities of all bodies immediately after impact. The impact is assumed to be fully elastic with the pulse being normal to the tangent plane at the impact point. The system as shown in Fig. 4 has no tree-structure. To arrive at a system with tree-structure one hinge is cut open.Al-

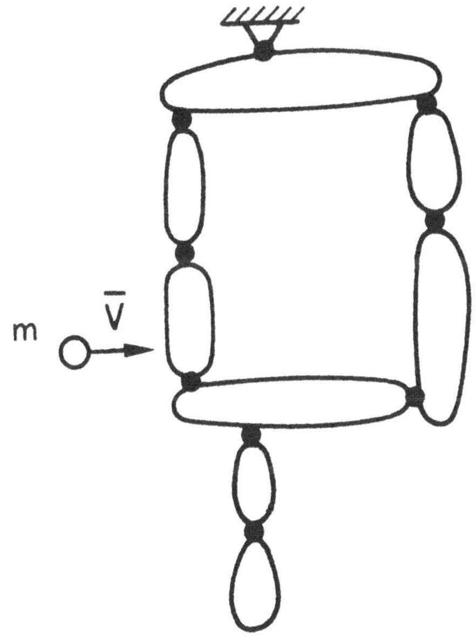

Fig. 4

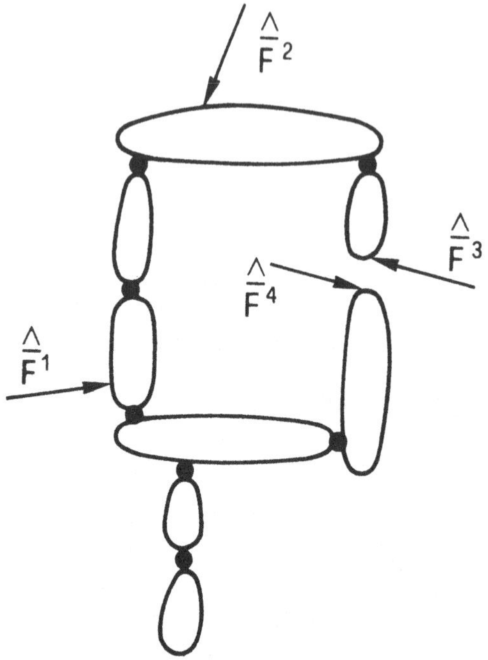

Fig. 5

so, the support is removed. The result is the system shown in Fig. 5 with unknown reaction pulses applied at the cutting points. For this system (15) combines twelve scalar equations with twelve unknown pulse components and twelve unknown velocity change components. The missing twelve equations are the following ones. The principle of actio = reactio requires that

(16) $$\underline{\hat{F}}^4_{abs} = -\underline{\hat{F}}^3_{abs} \qquad \text{(three equations)}$$

Kinematic compatibility requires that

(17)
$$\underline{\Delta v}^2_{abs} = 0 \qquad \text{(three equations)}$$
$$\underline{\Delta v}^4_{abs} = \underline{\Delta v}^3_{abs} \qquad \text{(three equations)}$$

Let the tangent plane at the impact point be defined by three mutually perpendicular unit vectors $\vec{t}_1, \vec{t}_2, \vec{t}_3$ with $\vec{t}_3$ being normal to the plane and let $\underline{t}_1, \underline{t}_2, \underline{t}_3$ be their component representations (measured in $\underline{\vec{x}}$). Then the equations hold

$$\underline{t}_1^T \hat{\underline{F}}^1_{abs} = \underline{t}_2^T \underline{F}^1_{abs} = 0 \quad \text{(two equations)} \tag{18}$$

A third equation is obtained from the fact that the component of the relative velocity at the impact point in the direction of $\vec{t}_3$ after impact is opposite equal to the same quantity before impact. Since the point mass m is subject to $-\hat{\underline{F}}^1$ this means

$$\underline{t}_3^T (\underline{v}_{abs} - \frac{1}{m} \hat{\underline{F}}^1_{abs} - \underline{\underline{\Delta v}}^1_{abs} = -\underline{t}_3^T \underline{v}_{abs}$$

or

$$t_3^T (m \underline{\underline{\Delta v}}^1_{abs} + \hat{\underline{F}}^1_{abs}) = 2m \underline{t}_3^T \underline{v}_{abs} \tag{19}$$

where $\underline{v}_{abs}$ contains the given components of the initial velocity of the point mass. The equations (16) to (19) together with (15) allow to calculate all elements of the column matrices $\underline{\underline{\Delta v}}_{abs}$ and $\hat{\underline{F}}_{abs}$. Subsequently, the desired angular velocity increments are obtained from Eq. (9) which reads

$$\underline{\underline{\Delta \omega}} = (\underline{K} + \underline{\Psi})^{-1} \underline{V} \hat{\underline{F}}_{abs} \, .$$

REFERENCES

1. Fischer, O., Theoretische Grundlagen für eine Mechanik der lebenden Körper, Leipzig, Teubner Verlag, 1906.

2. Wittenburg, J., Gleichgewichtslagen von Vielkörper-Satellitensystemen, Diss., Abhandlungen der Braunschweigischen Wissenschaftlichen Gesellschaft, XX 1968, 198-278

3. Wittenburg, J., Die Differentialgleichungen der Bewegung für eine Klasse von Systemen starrer Körper im Gravitationsfeld, Ing.-Arch. 37, 1968, 221-242

4. Wittenburg, J., Die numerische Bestimmung stabiler Gleichgewichtslagen von Vielkörper-Satellitensystemen, Ing.-Arch. 39, 1970, 201-208

5. Wittenburg, J., The Dynamics of Systems of Coupled Rigid Bodies. A New General Formalism with Applications, CIME Short Course on Stereo-Mechanics, Bressanone, June 1971

6. Wittenburg, J. Stossvorgänge in räumlichen Mechanismen. Eine Analogie zwischen Kreiseldynamik und Elastostatik, to be published

7. Wittenburg, J., Gleichgewichtslagen, Stabilität und Librationsschranken von Vielkörper-Satelliten, to be published (presented under the title "Bounds on Librations for Multi-Body Satellites" at the 21st IAF Congr., Konstanz, 1970)

Page

Page

Printed in the United States
By Bookmasters